PENGUIN BOOKS

MAD DOGS AND AN ENGLISHWOMAN

Crystal Rogers was born in 1906, and spent her childhood in India. Having moved back to England with her parents, she served as an ambulance driver and in a mobile canteen and library during the Second World War. After spending a few years in South Africa and Australia, Rogers arrived in Delhi in 1958. Here she opened an animal shelter named The Animals' Friend, which she ran for twenty years. She then set up a charitable trust named Help in Suffering in Jaipur. Rogers moved to Bangalore in 1990 and founded Compassion Unlimited Plus Action (CUPA), another animal welfare organization. She died on 30 August 1996.

Mad Dogs and an Englishwoman

The Memoirs of

Crystal Rogers

PENGUIN BOOKS

Penguin Books India (P) Ltd., 11 Community Centre, Panchsheel Park, New Delhi 110 017, India
Penguin Books Ltd., 27 Wrights Lane, London W8 5TZ, UK
Penguin Putnam Inc., 375 Hudson Street, New York, NY 10014, USA
Penguin Books Australia Ltd., Ringwood, Victoria, Australia
Penguin Books Canada Ltd., 10 Alcorn Avenue, Suite 300, Toronto, Ontario M4V 3B2, Canada
Penguin Books (NZ) Ltd., Cnr Rosedale & Airborne Roads, Albany, Auckland, New Zealand

First published by Penguin Books India 2000

Copyright © Compassion Unlimited Plus Action 2000

All royalties and proceeds earned from this book will be used solely for the welfare of animals.

All rights reserved

10 9 8 7 6 5 4 3 2 1

Typeset in Garamond by Digital Technologies and Printing Solutions, New Delhi

Printed at Chaman Offset Printers, New Delhi

This book is sold subject to the condition that it shall not, by way of trade or otherwise, be lent, resold, hired out, or otherwise circulated without the publisher's prior written consent in any form of binding or cover other than that in which it is published and without a similar condition including this condition being imposed on the subsequent purchaser and without limiting the rights under copyright reserved above, no part of this publication may be reproduced, stored in or introduced into a retrieval system, or transmitted in any form or by any means (electronic, mechanical, photocopying, recording or otherwise), without the prior written permission of both the copyright owner and the above-mentioned publisher of this book.

This book is dedicated to
all the animals who shaped Crystal Rogers*'s life*

Contents

Acknowledgements

Our grateful thanks are due to Penguin India, our publishers. Special thanks are due to Rosemary Poole, CUPA's UK representative. Without her, this book might never have seen the light of day. Our special thanks also to Sally Knocker, for her support.

We are grateful to Mrs Maneka Gandhi for all the support she has given to Crystal Rogers's work in the form of assistance not only to CUPA but also to all animal welfare organizations in the country. Our gratitude and deep appreciation also goes to D.R. Mehta who, during his tenure as Secretary to the state government of Rajasthan in the early 1980s, knew Crystal well. CUPA is indebted to him for his encouragement and support. Our thanks to Vivek Menon in Delhi, for strengthening our belief that Crystal's story was indeed worth publishing for animal lovers all over the world. CUPA would also like to thank Jacaranda Press, Publishing Consultants, Bangalore: in particular, Jayapriya Vasudevan for recognizing the potential in Crystal's story and her writings and for enthusiastically representing the book to Penguin; Amrita Chak for the massive task of consolidating Crystal's scattered papers, editing and giving definition to these memoirs, and Deepthi Talwar for her help at the final stage.

Preface and Disclaimer

Crystal Rogers passed away in 1996, leaving her literary estate to CUPA, the animal welfare organization she founded in Bangalore in 1991. CUPA (Compassion Unlimited Plus Action) commissioned this book as a series of vignettes based on a diversity of material—Ms Rogers's personal diary, the draft manuscript *Battle for the Weak*, letters, poems, essays, inter alia.

While compiling the memoirs and structuring the book, all care has been taken to ensure that the integrity of the writing is maintained. Events have been presented without alteration, as they were recorded by Ms Rogers. However, incidents and dialogues have been dramatized to enhance the readability of the book, while not distorting the original substance. Skeletal thoughts have been fleshed out while preserving the style and individuality of the author's writing. Where the author has not supplied dates and names of places, the endeavour has been to estimate chronology and location in context of the material. Spellings of people's names and the names of geographical areas appear as in the original material.

Prologue

'All My Travels'
—Crystal Rogers
18 March 1923

I was born in Bedford on 17 November 1906; the fourth child of my parents, I came at the end of a line of brothers. My father, Colonel G.W. Rogers DSO, was with the Indian Army and raised the 2nd Battalion 1st Gorkha Rifles in 1886. Most of the men in my family from both sides had served in the Indian Army. By the time I was born, Dad had retired, but he always wanted to return to India.

The following summer, I am told, we went to Southsea. We stayed in lodgings quite near the sea. Mother says this was a very picturesque place.

*

I was four when we went to Neuchatel, Switzerland. Andrew came to stay with us and spent six weeks—we used to go out boating a lot with Andrew rowing. In the beginning of July we went to St Croix, a little village in the Jura Mountains. In Saanen Moser I remember very well going out to pick berries and thinking to myself that this was a fairyland—listening to the sheep bells coming down the mountainside as I watched from our balcony.

There was a large ballroom in the hotel where we were residing. It had polished wooden floors. They were very slippery. Another little girl and I used to slide and pretend to skate. This same girl, I remember clearly, used to go around killing butterflies. She said they ate clothes. I never liked her very much after this.

By September it was time to come back home to England. We went to Folkestone where I had my fifth birthday. In the room next to ours there was a dear Irish terrier called Paddy. He was a great friend of mine. Sidney, the son of the landlady, was very kind to me. He gave me a book. One day in the greenhouse I found a boat which Dad and I started taking out to sea. We used to sail together and we would also go out on long walks. These walks I remember well—Dad would talk about so many interesting things. About the world and its people, about jungles and animals, about caring and being cared for.

*

By end-September we were in London, staying at the Grenville Hotel. Mrs Rook and Mrs Archer were very kind to me. They used to take me out in their hired car when they went shopping.

That year, on Christmas day I went to my first pantomime—*Cinderella.* I was not in bed until one o'clock! The manageress of the hotel gave me a Christmas tree and a musical box.

*

Dad used to take me out a lot. I used to go on long rides with him in the bus. Dad sailed for India on 16 December 1912, as he could not stand the cold. The wet climate of England did not suit him at all and he wanted to return to the Indian sunshine. Mother and I sailed in January. By now I was six years old. I thoroughly enjoyed the voyage. One of my best friends on the ship was a sailor called Percy who used to tell me long stories. There was also a Major Flemming who was kind enough to give me a necklace and two bracelets.

We arrived in Bombay on 7 February and set off straight away for Jellundar. We stayed with Uncle Phillip and Aunt Jess. Georgie and Neville came to visit and, of course, Dad was already there. Then we went up to Kashmere. We lived on a houseboat and travelled for two months up and down the Jelum River. Later we went up to live in a hut in Gulmarg. Colonel Rose drove us in his car.

I had many friends in Gulmarg. It was full of children and there were always a great many parties. I had a pony called Blacky and would ride everywhere on his back. Back in Sirinagar we stayed with Mrs Walton. Georgie came and stayed with us for six days. He gave me a dog called Bob. One night when we were in camp in Munshi Bagh, Bob saved my tent from being burgled.

*

By the winter, we had shifted back to Jellundar. Miss Holmes, my governess, who had been with us all this while in Kashmere, took leave and went to join her sister in Lahore. I had my seventh birthday in Jellundar and just about then Neville was appointed to 2nd 4th Gorkhas and had to leave for Bakhloh.

We went to Dharamsala and had a ripping house above the parade ground. I learnt music from Mrs Wishman since Dad was very anxious that I should be good at music. We didn't stay long here and soon enough we were travelling again, this time to Amritsa. I remember my eighth birthday in Amritsa spent with a cold in bed! Dad was ill. From Amritsa we went to Agra where we stayed at the Hotel Metropole. By now Dad was very ill. He was in hospital for three weeks. Neville came to spend ten days with us in answer to a telegram from Mother. But, luckily, Dad recovered and grew stronger and we decided to take him to Coonoor in the Nilgaries to stay with Miss Wells who used to take paying guests. Once in the woods adjoining her garden, I saw an enormous snake.

I used to spend as much time as I could at the roller-skating rink at the bottom of our road. This was a tremendous joy to me.

*

In November, Mother, Dad, Arthur and I all went to Bangalore where we stayed at the West End Hotel. I had my ninth birthday there. Our rooms were in a little annexe and we were awfully comfortable. I remember a young Scotsman who used to take me for rides on his motor bike.

I had a parrot and a white guinea pig called Mr Macdonald—not named after the Scotsman though!

There was a dear old storekeeper who, when we were ready to leave in March, gave me a baby duckling that had been hatched by an old hen. I called it Tommy Atkins and it turned into a huge Muscovy duck and a great character to boot!

We returned to Coonoor. The whole time that we were there Dad was slowly getting worse. Some time later came the terrible news that Neville was wounded. He was sent to hospital in Bombay. Mother went down to see him. She was with him for about ten days after which he was sent to a hospital in England. When Mother returned to Coonoor she discovered that Dad was dying. He died on the twenty-seventh day of April.

After his death we went for a while to a hotel in Ootacamund. While there, Mother saw one day in the paper that Neville had died of wounds at sea. Two days before Dad. The telegram giving us this news never reached us.

Early in June, Mother and I left for Dunga Gulli. This was such a lonely time for us …

*

I spent about four years in India, from the age of six to ten. I loved those years. We travelled around a lot, met lots of people and saw some beautiful places. But we lost Dad and Neville and that was so heart-breaking. I didn't go to school but was instead instructed by a governess. My mother wanted me to grow up as 'nature's child'. She used to tell me, 'I love to see you run like the wind. I like to watch you climb the trees and chatter with the monkeys. I should always like to watch

you caring for nature. Nature will in turn care for you.'
Sometime after Dad died we moved back to England.
First to Southbourne and later to Farnham in Surrey.
But life was always lonely without him ...

*

This is where it trails off. The ink gets smudged on the wrinkled paper. The preceding pages were written by me at the age of seventeen in the year 1923.

Now, at the grand old age of eighty-nine, I find it difficult to recall events from my childhood. I am grateful for this sketchy record which has endured many, many years of damp and oppression in a tin trunk.

Much happens in eighty-odd years: all kinds of little and big calamities; big achievements and small ones; many joys and much pain. But to document all of it is a task beyond my patience and ability ...

My early adulthood was haphazard—a variety of jobs and occupations came in a procession; some interesting, others lacklustre. It was not until I met a certain special person during World War II that I found direction. That direction ultimately helped me find a purpose. All at once, after years of floating like a rudderless ship, I had found a reason to be, to work and to live.

From here on, after this brief and decidedly incomplete introduction to my early youth, I will dwell only on selected moments of my life—capsules of time that have been unique and precious for me.

One

ALEXANDER JAMES PEPPER

My story begins in the very depressing years of 1941 and 1942. I had started the war as an ambulance driver. Later, finding the inactivity boring, I changed over to a mobile canteen and library on the east coast.

I was based in a village rectory, now turned into a canteen for the Ack-Ack troops and a neighbouring Royal Air Force base. The rectory was old and gloomy. During some of the long winter evenings I would play table-tennis with the men; other evenings would be spent mending my torn library books in front of the big kitchen fire, listening to the sound of heavy boots clattering up and down the carpetless stairs.

I tried to shake myself out of loneliness and depression by talking to the men, most of them good-hearted lads from the midlands. I liked the 'boys' and I think they liked me, but when it came to talking to them of the things that really mattered to me, it was more difficult than trying to converse with deaf mules in a foreign language!

It must have been some time during the first week of January 1942, that Miss Brawn, the grand old lady in charge of the canteen, came to me with a very ordinary request. 'There's a lad in the games room looking for a book to read. Have you got anything you could lend him?'

'I expect so,' I said. 'Have you any idea what he wants?'

'No. I forgot to ask him,' said Miss Brawn. 'The usual blood and thunder, I expect. You'll be able to recognize him,' she added as she went out, 'he's a Canadian.'

With a book or two under my arms I went down to the games room. Against the mantlepiece lounged a long-legged RAF flight sergeant with very curly hair and a flash on his shoulder which proclaimed him a Canadian.

'Hullo,' I said, 'I hear you are looking for something to read. What sort of books do you like?'

'Oh, I don't know.' His smile showed a gleam of very white teeth. 'I read anything. I guess I like open-air things best. A Zane Grey would do.'

'Most of my Zane Greys are out, but I have got a couple in the kitchen which I am mending. You'd better come and see if you have read either of them.'

The long-legged young man followed me to the kitchen.

'I say!' he exclaimed, as he saw the fire roaring in the grate. 'It's a lot cozier in here, isn't it? More like home!'

While I hunted for the books, he talked. He told me that he had arrived at the RAF camp just before Christmas. He was the only Canadian there, and everything seemed a bit strange to him. Obviously he was missing his home very badly.

As he continued to talk, I found myself watching him with increased interest. He had one of the most animated faces I had ever seen. He had nice grey-blue eyes that looked you straight in the face. He had deep laugh-lines on either side of his mobile mouth. When he laughed, which he did only too often, his eyes would suddenly screw up and almost disappear! He kept his light brown, curly hair neatly brushed back away from his high, broad forehead. One thing impressed itself on me from the very start: the feeling of cleanliness he brought with him. Not only of body, but of mind as well.

'What's your name?' I asked.

'Call me Jim,' he replied promptly. 'Otherwise I am Flight Sergeant Pepper . . . at your service.'

He got up and bowed and we shook hands. I noticed his were nicely shaped, with long, artistic fingers and well-kept nails.

'Jim?' I said enquiringly. 'Does that mean your name is James?'

'If you want the whole works,' he laughed again, showing his dazzling teeth, 'it's Alexander James Pepper. Bit of a mouthful, isn't it? At home they call me Alec, but since I joined the Forces it's been "Jim".'

'Right then. That is what I shall call you, Jim.' I handed him the books but noticed that he appeared reluctant to leave.

'I suppose you wouldn't feel like teaching me table-tennis, would you?' he asked doubtfully.

'Teaching you?' I asked, surprised.

'Well, I've hardly ever played, you see, and I am as fearful as a rabbit. I hear you are so darned good.' He was almost apologetic about the whole thing. I promised to play with him the following evening if he was free and had no better way to entertain himself. I was almost convinced that he would most certainly forget by the end of the day.

The next evening, however, I was the one to have forgotten my 'date', when somebody told me that a RAF bloke by the name of Jim was waiting for me in the games room. I hurried upstairs with many apologies.

For one who was a so-called novice at the game he moved with grace and was light on his feet. I decided that whatever Jim put his mind to he would do well. He took his beating in the best of spirits, and booked me for another game the following evening.

That was the beginning. I began to look forward to my nightly table-tennis with this immensely likeable young man, with his spontaneous laughter and a musical quality to his voice.

As I got to know Jim better, I also found that he was, in fact, very musical. I also found that he had a deeply religious side, though this was not confined to any particular branch of the Christian church. On one or two occasions we attended evening service in the village church together. I noticed that he did not even glance into the hymn book. He knew all of them by heart. Later, he informed me that 'back home' he had at one time sung in the choir.

I had only known Jim for matter of days when Miss Brawn started classes for those of us who wished to learn German. To my surprise, Jim signed up. The classes were held in the kitchen. While I sat mending my books, I could watch and hear Jim tying himself into knots with his pronunciation of German words. It was plain from the start that he did not intend taking these classes seriously. His tremendous sense of fun finally broke down even Miss Brawn's defences and she was forced to laugh with him.

The weather grew increasingly cold and finally the snow came. All Air Force personnel were grounded and for several days I was unable to go out with my mobile library. Jim was over daily and his table-tennis improved by leaps and bounds.

Some days later when the roads opened up again, Jim was overcome with shyness as he stood next to me watching another couple playing table-tennis. Finally, hesitatingly, he brought out his request: 'Tomorrow's Saturday. I don't suppose you would like to come into town with me?'

I looked at him, too shocked to speak. There were plenty of WAAFs at his camp. Surely this boy, at least ten years my junior, had someone nearer his own age to take out?

I hesitated.

'Please,' he begged, 'I thought it would be nice to go to the pictures, but it's so dull going alone.'

And then, suddenly, I thought I understood. Had he not talked again to me of his mother—and what great friends they were? How much they had always done together? Of course. He was just missing her now, and was in some way putting me in her place.

'Indeed, I shall come—if you haven't got anyone more exciting to go with. I'd love to. It would be fun.'

Seldom, if ever, had I enjoyed a date more. Jim's tremendous capacity for enjoyment was infectious. He had an extraordinary knack of getting more happiness out of life than anyone I had ever known. I have quite forgotten what film we saw. But I will never forget that afternoon. We had hot dogs and coffee afterwards in one of the canteens in town. I was in a merry mood and shared with him the reason why I wasn't married as yet. I recited a poem for him which I had composed some time ago.

Do you rise with the lark to welcome the dawn
And bare-footed walk through the dew?
Composing sweet sonnets and verses of love
To some maiden devoted to you?
Or do you come down to your breakfast each day
With a face like a rather wet week,
And prop up your paper 'twixt you and your wife
Whom you bite if she ventures to speak?

Do you wander inspired through the sun-dappled
woods
And bathe in the river each day?
Or race with the winds through the star-spangled fields
To join shepherds and nymphs at their play?
Or is your life spent at some ledger or desk
Where you work all the dreary week through,
And the only occasion where speed wings your feet
Is when you're catching the 5:52?

At the close of each day do you view with delight
The sun's setting rays through the trees?
And stroll in the twilight composing sweet odes
While your hair is caressed by the breeze?

Or in tightly packed trains to a strap do you cling
Like the monkey that swings from a tree?
Your umbrella well furled, and your bowler hat
perched
On the place where your hair ought to be?

Oh cherish illusion, let your fancy be free,
Inspiration continue to flow;
If you are not the man I should like you to be
What luck that I never shall know!

Jim laughed quite helplessly. He found it funny in parts and in others he said that it was very real. That most husbands must in fact be crashing bores. I agreed. When we had finished with our snack and were sitting in one of the rather bare recreation rooms at the canteen, Jim suddenly opened up a bit of his heart to me.

I understood something of his difficulties and temptations. 'I don't know what it is,' he said as he finished. 'I like 'em all and I guess they are a fine lot of guys—but I just don't want to do as they do. I don't seem to feel as they do. And I guess somehow it's just got to stay that way. But I'm dead scared of being thought different. I don't want the boys to think I'm toffee-nosed and too darned stuck up to join in their fun. But sometimes it's so darned difficult, I just don't know what to do.'

He stared down at his boots, for the first time avoiding my eyes, and continued, 'If you'll just stay by me and always be my friend I know that it will be all right. There's nothing I can't do with your help.'

I was deeply touched. I said, almost in a whisper, 'I will always stay by you.'

Two

FIRST LOVE

From that afternoon there was a feeling of closeness between us which I find difficult to put into words. We somehow understood each other without the necessity for speech.

By now we had discovered our common love of music. Jim was perpetually persuading me to play the piano at the rectory. When nobody was about I used to get the music scores of my songs and Jim would sing with me, his tuneful baritone voice harmonizing with mine. Of all the songs we sang his favorite was *I'll Walk Beside You.* He would want to sing it again and again.

'I like the tune,' he said, 'but mostly I like the words.'

Sung by him the words lost their commercial sentimentality. He sang them as though he meant them; as though they carried some special promise intended only for me. I tried to brush the thought aside as imagination. I told myself I must remember my age and try not to take him too seriously.

Jim was still grounded when I started again on my rounds. I got back to the canteen that first evening, my feet frozen with cold. Jim was waiting. I had hardly mentioned my cold feet before he threw off his fur-lined flying boots.

'Here—put these on!' he cried, then doubled up with laughter as I almost disappeared inside them. 'I can't see why you can't wear 'em,' he protested, 'even if they are a bit on the big side. I guess it's doing them no good just sitting around while I'm grounded.'

The grounding of the RAF was not destined to last much longer. Within a few days a series of devastating attacks were being carried out over Berlin and Hamburg. Jim never wanted me to know when he was going out on 'ops', but I invariably managed to find out. It was on one of these evenings that I pressed a tiny horseshoe into his hand.

'To bring you good luck,' I whispered.

'I'll always keep it,' he whispered back, and was gone.

*

It was soon after the worst raids started that I got a transfer. I had asked for a move before Jim came upon the scene but had not expected it to be so soon. I was now to be based at B——. There were many points in its favour, but suddenly I found that I did not want to leave.

I broke the news to Jim as soon as possible. At first his face fell. Then he cheered up and said, 'They might have sent you a lot further off. You know I'll be over the minute I'm off-duty.'

He insisted on coming with me to settle me in. A back-street small house had been allotted to me for stores and so on, but I myself was expected to live in digs. I had other ideas, however, and outlined them to Jim.

'It's an absolute waste to live in digs with that little house going empty,' I said. 'It's got some basic furniture. With a few licks of paint it could be made quite nice. One of the rooms could even be turned into a home away from home for the boys, with newspapers, magazines, a gramophone and things,' I said.

'It's a wizard idea!' he said enthusiastically. 'I'll come over and help you paint and we'll make an absolute palace of it before we've finished!'

Gradually I settled into the daily routine. Jim came over every day when he was not on duty and there were long evenings when we would sit in front of the fire and he would talk of 'home'.

His mother figured prominently in most of his reminiscences. It seemed he had a special place in his heart for her. She was a constant figure in his life and in his conversations with me.

He was the youngest of his family by a good many years and was still at home when his brother and sisters had gone out into the world. It was a big wrench for him—parting with his mother and leaving her alone. His father, a sea captain, had died when Jim was only a few years old. Jim's first civilian job had been as a radio operator in a small radio station in the far north. There, his dwelling had been a crude hut; his only means of transport, a sledge over the snow. He loved the free open life but with the war in Europe he felt the urge to join up. Everything had been done to persuade him to stay. But the urge had been too strong and he volunteered for the Air

Force. After completing his training in America, he had been drafted to England and had joined the Royal Air Force. He arrived at his present base immediately upon getting his promotion.

He had stopped his German lessons after I had left for B——, and did not seem to have any further interest in the subject.

'You don't seem to have got very far with your German,' I remarked one day. 'What on earth made you take all those lessons?'

His eyes opened wide in surprise.

'Why, only to be near you,' he said simply. 'No other reason.'

Though his official job was that of radio operator, a shortage of personnel had given him a changeover and he was now acting as air gunner.

He brought over his painting overalls, but somehow we never got down to the painting. The night raids were continuing and were taking their toll on his vitality and spirits. His irrepressible sense of the ridiculous would break out whenever he could find anything to laugh at and his best jokes were always directed against himself. I never heard him speak unkindly of another soul, but slowly I was beginning to see more and more of his serious side. Once, after a big raid over Hamburg, he almost broke down.

'It's ghastly!' he cried, in a voice which he struggled to keep steady. 'Just hell let loose! To look down below and see everything in flames, and to know that there are women and children down there . . .' He choked and buried his head in his hands. 'I'd be no good as a soldier,' he went on after a little while. 'I just couldn't kill anyone however hard I tried.'

About this time Jim's headaches started. He would come over to my little cottage, apologize for a blinding headache and just slump down on my sofa and close his eyes.

'Please, dear, could you just sit beside me and hold my hand? Then I shall feel all right. I'm so sorry to be like this,' he would apologize.

Drawing up a chair beside him I would sit holding his hand until the tense lines disappeared from his face and he drifted into sleep, looking like an exhausted little boy. What was this extraordinary attraction I had towards this man, younger to me by so many years; what was this extreme tenderness I felt for him? This was an emotion I could deny no longer.

Oh, God, keep him safe always, I would pray often. I was beginning to feel a bit afraid of this obvious devotion he had for me. He was making no attempts to hide it. It was puppy-love, I told myself firmly, or calf-love, anything you like—but it certainly could not be anything more. For his own sake, I told myself, I must not encourage him in any way. I must not let him see how much he mattered to me. But all the time it was becoming more difficult . . .

We had so much in common—our love of nature and of animals, of music and of travel. Jim had a longing to see more of the world. He would recite the names of the countries and end it off with: 'But I should always like to have you by my side, of course!'

I would smile to myself and marvel that he should

always want to include me in his dreams of the future. With his friendliness and abounding joy in life, he made friends easily enough. Any girl he met could hardly have withstood his charms and would gladly have taken him for the asking. It never ceased to amaze me that he apparently preferred my company to theirs.

'People are just so darned nice to me!' he would exclaim sometimes in delighted surprise. 'I guess I don't know what I've done that folks should always be so swell to me, the way they are.'

Jim also had very strong ideas on ethics.

'I can't understand Will,' he said, naming a Canadian friend of his. 'He's only been over three weeks and already he's got a girlfriend and is threatening to get engaged to her. It's all wrong. A fellow's got no right to get engaged when there's a war on. It isn't fair on the girl.'

I went over to the rectory several times on my day off. Jim was invariably there to meet me, and as the weather got warmer we would walk together through the countryside. His headaches were still blinding him and on one occasion he had a complete blackout for several minutes while on operations. Once, when we were out for a walk together, he told me he was beginning to feel the strain of the night raids.

'Do you mind if we don't go too far today?' he asked, flopping down on the grass on the hillside. I sat down beside him. He stretched himself to full length and slipped his hand into mine.

'Do you mind?' he again asked—but before I could answer him, he was fast asleep.

Again I was struck by his look of rare purity. He was so utterly guileless. He couldn't hurt a fly and bitterly I rebelled against the fate which took him night after night into that frightful hell of death and destruction which he had described so vividly to me. Suddenly a cold breeze sprang up and with it a cold feeling of fear entered my heart. I shivered and Jim awoke, cursing himself for wasting my afternoon.

A few days later Jim got leave to visit an aunt and uncle up in the north of England.

'I don't want to go,' he said. 'I'll hate leaving you. But I promised to go, so I must.'

I laughed at him. 'Don't be silly! A rest is just what you need. You must get rid of those headaches, and you've started a nasty cough, which I don't like either. A change will do you good and you'll come back all the better for it.'

I then put forward a proposition that had recently entered my head. A cousin of mine, who had a house upon the Thames near Oxford, had told me that I was welcome to come down any time I got leave, and if I cared, to bring a friend. I wondered if Jim would care to be that friend. He had told me that he had more leave due to him and if I could manage to make mine coincide with his, we could visit my cousin's house and have a restful holiday on the river.

Jim was thrilled with the prospect and it did much to cheer him as he set off, reluctantly, for the north. I received a letter from him every day that he was away. I was delighted to know that he was safe and was resting. And it was wonderful to know that I was being missed!

When he returned, his cough was a bit better but his headaches were still bothering him. This worried me. I wanted him to be examined by the medical officer but he kept resisting the idea.

'They'll surely ground me if they get to know, and right now every man is needed.'

I had only to bring the matter up for him to change the subject immediately: the one that claimed most of his interest was our proposed visit to my cousin's house. He was just like a child waiting for some special treat and had even begun to decide on how many white shirts to take with him and the importance of having plenty of clean collars!

It had become a habit of Jim's to phone my canteen every morning with a message to say that he was all right. One morning I received no message. No fewer than eleven of our planes had been brought down that night. I was in panic. The whole morning the words kept ringing in my ears: 'Eleven of our planes are missing! Eleven of our planes are missing!'

At lunchtime, unable to bear the suspense and torture any longer, I rang up the RAF Sergeants' Mess. Could I speak to Flight Sergeant Pepper, please?' I asked. After a long pause an orderly came back with the news that he could not be found.

'When did you see him last?' I managed to whisper, hardly able to speak the words.

The orderly was quite indifferent.

'Nobody's seen him this morning,' he said.

How I got through that afternoon's round, I don't

know. I did what I had to do in a mechanical daze. 'Eleven of our planes are missing,' said the wheels of my van as I drove along the bleak east coast roads.

At last I got back. Hardly able to see what I was doing, I stumbled through the doorway of my little back-street cottage. Then I heard a light footstep behind me and turned to find Jim standing there with his usual infectious grin.

'Jim!' I almost sobbed, and the next moment his arms were around me and I was clinging to him as though I never meant to let him go. Floundering, I told him the whole story—my telephone call, and then my fears that he had not come back.

'You poor kid! You poor kid!' he kept repeating tenderly as he continued to hold me tightly. 'It's all my fault. The guy who usually wakes me forgot to and I just overslept. But it's all wrong. I shouldn't have let you …'

'Shouldn't have let me what?' I asked.

'Shouldn't have let you feel this way about me,' he replied seriously.

I gave a shaky laugh. 'You silly chump! As though you could have possibly helped it!' I said. Then, once again, overcome by his nearness and how precious he was to me, I added, 'Don't ever not come back, Jim.'

He looked at me tenderly and said, 'I'll always come back, darling. I promise.'

Before we parted late that evening, he assured me that he would see the medical officer about his headaches.

'Don't you see, Jim,' I persuaded him, 'if you have another of those blackouts like you had recently, it might put everyone on your plane in danger. You have got to get yourself put right if you are to do the job properly. Besides I don't want to take an invalid with me to Oxford!' For once Jim saw the wisdom of my words.

'All right, honey—I'll see the doc,' he said. 'He's bound to ground me for a few days, so I'll come over the day after tomorrow and we'll start painting. We've put it off quite long enough. What's the good of my having brought my overalls here if we don't do any painting?'

'Not the day after tomorrow, Jim dear,' I said. 'You've told me several times that Will is getting a bit restive because you haven't been out with him since he came to the camp. Since he's an old friend from home, I do think he ought to have a little of your time. Do something with Will on Friday and with me on Sunday.'

Jim made a small grimace.

'I'll see if Will wants to, first,' he said. 'He may be wanting to go off somewhere with that girlfriend of his.'

I laughed. 'All right. I'll expect you if I see you, but not otherwise.'

And so we parted—I light-hearted for once at the thought that for a few days at least I should be able to go to bed without the awful knowledge that Jim was somewhere out in the darkness, his mind in torture and his life in danger.

For the next two nights I slept soundly, happy in the knowledge that Jim was safe. For some reason I felt that the crisis had passed and that Jim had somehow proved to me that he would come through all right and that nothing, henceforth, could harm him. On that Friday I went through the day completely happy with my work and constantly breaking into song.

I got back to the canteen about 6 p.m. to find a message waiting for me. A young RAF sergeant had come in with a friend apparently and had left a message for me to ring him up at the local cinema as soon as I was free.

'Jim,' I thought happily. 'He's evidently got Will with him. Why should I butt in on their party? It's all very well for Jim, but Will may not want it,' and I refrained from ringing.

Later in the evening when I visited the canteen again, I was told that I was wanted on the telephone. I went to it cheerfully.

'Hullo, Jim,' I said.

'It's not Jim. It's Will,' came the reply.

'Well, hullo, Will!' I cried. Then suddenly a foreboding came over me. 'Is Jim all right?' I asked apprehensively.

There was a pause.

'Well, actually that's what I'm ringing you up about,' said Will. 'You see, he was sent out on ops last night … and … he's not back yet.'

Three

WALK BESIDE ME

It was two days after I had heard those fateful words from Will that I received a letter in Jim's own rounded handwriting. At first I could scarcely believe my eyes, but on tearing it open I discovered that it had been written on the morning after I had seen him last. In the letter he wrote that he had seen the medical officer and reported his headaches. Since, on that particular day, he did not have a headache and the weather had turned warmer, the medical officer had suggested that it would do him no harm to go out on operations. He would, he promised, be seeing me on Friday anyway, as he was anxious to get on with the painting of the cottage. 'Don't worry,' the letter concluded, 'for when you do, I worry too and I know you wouldn't want that.'

There is no point in detailing my feelings during the weeks and months that followed the news that Jim was missing. For me the sun had gone behind a cloud. I do not think I had realized fully before, how completely Jim had altered my life. My loneliness had been magically routed by the brightness of his smile. In his joyful company it was quite impossible to feel anything but happy. My heart went out to his mother. If just a few short weeks of his company had produced the awful sense of loss which I now felt, what must she, who had brought him into this world, and known him all his life, be feeling? I wrote to her and a correspondence followed, lasting until her death eight years later.

Meanwhile, I continued my work mechanically, in a semi-dazed state. In my mind three words kept repeating themselves endlessly: Missing! Feared killed!

Poor Will! He too was feeling Jim's loss greatly. Although we had never met, he had both written and telephoned and promised to see me upon his return from leave. This promise, however, was not to be fulfilled. Upon his return he was sent on operations. Like Jim, Will too was never to come back.

I was very much alone. I had no real friends in the neighbourhood, no one I could turn to for comfort. The hours dragged by on leaden feet. There was not one minute that I was not thinking of Jim. Hoping, praying, wondering ...

During one of the dreary days that followed, I wrote a letter to Jim. I wrote to him as though I believed him to be alive. In it I told him all the things that I had never dared to say, but somehow I felt he had the right to know. It was a long letter and after it was written I put it away until such time as I could give it to him myself, although all the time I knew in my heart that he would never read it.

Three months dragged wearily past, till in July the official records came through at last. Jim was dead. Along with all the other occupants of his bomber which had come down somewhere in the neighborhood of Kiel in Germany. The news put an end to any hopes which lingered, but to me it was not really a shock. I had already known . . .

It was about this time that I snatched at the only straw of comfort I could envisage, visiting the Spiritualist churches in the neighborhood. I sat through uninspiring services where batty-looking mediums demonstrated their doubtful gifts of clairvoyance to credulous congregations. On more than one occasion I was given 'messages': once from a 'dear old lady with her hair parted in the middle and a nice oval face' who might have been, according to the medium, a relative from my mother's side. Upon another occasion I was assured that a small child in a white pinafore was putting a bunch of pansies in my hand! I was not in the least interested either in the dear old lady or in the pinafored girl with the pansies. I wanted one thing and that only. I turned disgustedly away from spiritualism; it was fit for imbeciles.

A short time later I was in Germany. Finding myself stationed not far from Kiel, I made it my business to find Jim's grave. This I could never have done without the kindly help of a member of the British War Graves Department who happened to be in Germany at the time. We found the grave at last. It was, however, in company with those of all the other bomber crew. There were no crosses and even the mounds above the graves were partly obliterated. I went back to the town where I was stationed and had an army carpenter make me six crosses, upon which I had the name of each man written and the simple inscription: 'Alive for ever more'.

It was the last thing I did before leaving Germany. I was staying at the time in a German hotel which had been commandeered for our use. It was just before I left that I was awakened one night by a sharp double rap on my door. Thinking that it was morning and that a colleague might be waking me, I sleepily called out, 'Come in!' Nobody came. I opened my eyes and looked at my watch to find it was two in the morning. Wide awake at once, I jumped out of bed and threw open the door. There was nobody there and the landing was completely empty. Thinking that someone must have mistaken my room for another, I went back to bed and thought about it no more. But when the same thing happened again at the same time the next morning, I was somewhat mystified.

By 1946 I was in England, my war work over. I was restless and unhappy. I didn't know what I wanted to do next. I had been back several months when a chance remark made me prick up my ears. It referred to a certain woman doctor, of whom I'd never heard before, and some amazing evidence she had given regarding someone else who had been killed in the war. My informant gave me her name and a way to reach her.

'Don't you know anyone who knows her?' I asked.

'Well, I believe the Lindseys are friends of hers,' she replied. 'Do you know them?'

I did, but not that well. However, I took my chances. I rang them up; Mrs Lindsey was obviously surprised to hear from me again, not having heard from me since the beginning of the war. I cut the preliminaries short and plunged straight into my request.

'I believe you know Dr Graif?' I said.

'Yes, certainly,' said Mrs Lindsey, 'she lives quite near us.'

'I'm awfully anxious to meet her,' I said.

A meeting was arranged for me the following afternoon. At four sharp, I arrived at Dr Graif's door. She had only come to live there recently and took me around her lovely garden before tea. Our talk had no bearing upon the subject I had come about. I still recollect very clearly that we had home-made scones for tea and we buttered them ourselves. I was just raising a scone to my mouth, and my hostess was pouring out tea, when she remarked in the most matter-of-fact voice: 'Who is the young airman who came in with you?'

She then went on to describe in vivid detail the appearance and personality of Jim. I couldn't believe my ears. From the complete naturalness of her description, Jim might have been standing beside me in the room!

I don't remember what I did or said. I had got just what I wanted, but now that it was happening, I hardly seemed to be able to take it in. It was almost too good to be true.

'He's terribly worked up and excited,' said Dr Graif. 'He says he's been trying to get through to you for so long, but you would bury yourself in a cloud of gloom where he couldn't reach you. He says his name is Tim or Jim. Something like that. He was killed in an aeroplane which came down and he was very closely linked with you.' She paused, then went on. 'He's in such an emotional state that it's difficult to follow him. His image keeps fading. If this is the first time he's got through, he'll probably find it much easier to make contact as time goes on.'

I don't remember a single contribution that I made to this conversation. I only remember that at the time I was convinced to the very core of my being that here, in this room, beside me, although I could not see him, was the loving, joyful being I had known as Jim. I felt myself enveloped in his warmth.

As I walked out of the house half an hour later, the sun which had seemed to have hidden itself from me for the past four years came out suddenly from behind the clouds. And I walked home bathed in its golden glory.

*

I wanted to believe more than anything else in the world that Jim still lived and that the link between us was not broken. But the more I wanted to believe this, the more difficult I found it.

My Sunday afternoon experience had shaken me to the very foundation. While at the time I had been completely convinced of Jim's presence, within a week I was doubting the whole thing. After all, I told myself, Dr Graif was quite obviously sensitive—there was nothing to prevent her drawing all that information from my own mind! She would no doubt be proficient in telepathy. Besides, she had not told me anything that I did not know already.

It was at about this time that I met the parents of the pilot who had flown Jim's plane. I had sent them a photograph of the graves in Germany. A correspondence had followed, in which they invited me to visit them upon my return to England. I went to see them one day at their home on Kingston Hill and within a few minutes of meeting them, discovered them to be spiritualists. They had endless proof, so they told me, that their son Bob was still living and happy, though in another plane of existence. They begged me to join the London Spiritualist Alliance, where, they assured me, I should get all the proof and comfort I needed. Despite my doubts, I found it difficult not to feel tempted. My link with Jim was deep and lasting; it belonged in a realm where time and age and material boundaries did not exist. I was feeling more and more compelled to search for him.

As a result of joining the London Spiritualist Alliance I made several new friends. Amongst these were two sisters, Mrs Stewart and Mrs Denton, who told me that they had been communicating with their husbands over a great many years. For this they did not employ a medium, but simply sat together with a table between them upon which were the letters of the alphabet and an upturned wineglass. On placing a hand each upon the base of the wineglass, it apparently moved about the polished surface of the table, stopping in front of letters and spelling out long messages.

Greatly intrigued, I accepted their invitation to a 'wineglass sitting'. They had hardly settled at the table before the wineglass spelled J-I-M. The sisters knew no one of this name and asked me if I did.

'Is it Jim for Crystal?' asked Mrs Stewart, who was taking down the notes.

'Yes, yes. Jim,' the wineglass spelled out rapidly, then continued, 'happy returns.'

This was hardly evidence, since the sisters already knew that it was my birthday. By way of a test, I asked what I had in my bag which was given by Jim? The wineglass spelled out 'photograph'. It was true. I did have a photograph but this was not the answer I had wanted. Shortly before he died, Jim had given me a tiny bottle of scent and this I had kept.

'A little p-' spelled out the wineglass in a hesitant manner, 'a little pe-' and once again it seemed to hesitate.

During this slow and lengthy spelling process, my mind was beginning to wander. I was worried about my little dog, wondering how much longer she could live.

Suddenly, Mrs Stewart, who as note-taker, was watching the letters, gave a little exclamation of surprise.

'Why! He's spelled out C-H-U-F-F-Y! Isn't that your dog? How absurd! How could you possibly keep a dog inside your bag?'

But in view of what I had just been thinking it was not quite that absurd. The wineglass, now, was spelling rapidly: 'Three ten. Three ten. Light. Light. Dog.'

It took some amount of thinking before we all decided that this might be reference to a monthly magazine called *Light*.

'Do you mean page three hundred and ten?' Mrs Stewart asked, and got the answer, 'Y-E-S.'

Frankly, I felt very doubtful about the whole thing. It was not till later that evening, when I l got back home with a copy of the magazine, that I discovered that page 'three ten' actually existed. To my further amazement this is what I read:

'The following curious incident is related by a Fellow of the London Spiritualist Alliance:

'For fourteen years I had a much loved Cairn terrier who never left my side day or night. At the end of last year I had to put him to sleep as he was in pain. On the twentieth of June this year, I was sitting reading in the car at Buckie, waiting for the time when a meeting was to begin. Two schoolboys, aged about ten, came up the street, stopped and looked at the seat beside me, and one said to the other, "Look! What a lovely little dog." This was the seat where Hamish always sat beside me with his head resting on my knee. I regretted later that I had not got out and spoken with the boys, but I was too stunned for some minutes after they had gone.'

I read through this account time and again, in utter amazement. Was I shortly to lose my small, beloved dog, Chuffy? Was this Jim's way of trying to bring me comfort? He was a great animal-lover and had been very fond of Chuffy.

Then gradually, my critical faculty reasserted itself. How could I be sure that Mrs Denton's subconscious mind had not made contact with mine and thus had recognized the fears that were uppermost in my mind at that time?

I considered again the abortive attempt to describe the article in my bag. 'A little p-. A little pe-.' If Mrs Denton's subconscious had been able to see other things in my mind, could she not have seen this? Suddenly light dawned! Jim had been a Canadian and to him 'scent' might very well have been 'perfume'. Perhaps he might have been trying to say: 'a little perfume bottle', but for some reason had been unable to get it across? The fact that Mrs Denton had not been able to establish the nature of the article in my bag suddenly seemed reassuring. I pondered over the whole question long and deeply. And despite all my doubts I found this entire episode comforting.

Within three weeks Chuffy breathed her last.

Four

FOREVER ALIVE

Several years passed by: I travelled to South Africa and Australia and even in these countries my quest for the perfect contact continued. There was much evidence piling up to convince me that life, in fact, did not end on earth, that the people we cherish are always walking by our side. Perhaps it was a fear that this belief may turn out to be fanciful that made me keep believing in the possibility that all such contact was coincidental.

At a psychic session, back in England, I got considerably more information. I was told repeatedly to cross the river. I asked, 'Which river?' And the answer was 'Thames'. This, however, was not very enlightening. I did not see how it would help my psychic development in any way.

I decided to start a home-circle of fellow-seekers. I put out a small notice asking for those interested to join me. A week later a couple contacted me. They asked me to visit their home. They lived just across the river Thames …

Well, I crossed the river. Jack and Claudia, my new acquaintances, decided to waste no time. We started that very evening without any preamble. There was no exchange of information, no pleasantries. A wineglass was brought out the letters of the alphabet placed in the middle of the table, and the session began. At the first attempt we got a lot of movement which was described as 'messages', but nothing that I could consider evidential. I left disappointed.

The very next day, as Jack and Claudia sat conducting a session by themselves, the name 'Alec' was spelt out for them. They did not make any connections with it and asked me if I did. I feigned ignorance. I could never be sure how much information mediums could rake up about the past histories of their clients or members of their circle. That night I sent out an earnest prayer to Jim. I asked him to give a description of his hair to Jack and Claudia. I said, 'Say it was curly, kinky, wavy, anything you like. Say you were a woolly-headed golliwog, but do try to say something about your hair.'

A few days after this, Claudia and Jack practised automatic writing together. They could get only one sentence. Meeting them the next day, I asked what success they had had with their writing.

'Absolutely none,' said Claudia. 'We couldn't get anything but rubbish last night. It was just a lot of scribbling and only one sentence: "Jim—Zulu man!"'

On another occasion, practising automatic writing with her husband, Claudia contacted someone who called himself Hugo. He professed to have known Jim. She asked him where they had known each other. The pencil immediately drew something that resembled an uneven star. This conveyed nothing at all to Jack and Claudia.

However, when they showed it to me, I had no difficulty in recognizing it as a roughly drawn maple leaf. Canada! Jack and Claudia had no idea that Jim had been a Canadian.

*

On subsequent occasions, Claudia got a barrage of information. She would be exhausted after these sessions. It was clear that Jim was extremely earnest about being heard. She would report back to me that she had seen, in fragments, an explosion. A feeling of falling and falling. Cold. Icy cold. An effort to escape. A huge desire to live. But at the end there would always be the assurance that the communicator was all right. He was safe and happy.

I have always believed that if Jim had given his medical officer an accurate account of his headaches, instead of making light of them, he would never have been sent out on that last fatal night.

By now I had been introduced to a new concept: the direct voice contact. This implies a communication where, either through a medium or through a trumpet, one receives messages from the other side that are spoken rather than written. It felt like a small step towards a more fulfilling experience. With Jack and Claudia I discovered that there was tremendous promise to these sessions if you found a genuine medium. I had satisfied myself that they knew nothing whatsoever about my past. I had been extra careful in not allowing any telling information to slip through during our interaction. Yet, I hankered for more.

*

It was now late summer, 1950. Eight years and more since I had heard those fateful words: Missing. Feared killed. For so many years evidence had been coming my way, but still the old bugbear, 'It might all be telepathy!' prevented me from believing. I was terribly afraid of deceiving myself. I went through many direct voice sessions, some disappointing, some uplifting. I was still to meet the perfect medium.

Somewhere around this time, I was introduced to a Mrs Blackridge. She took me in for a sitting immediately. This was to be the turning point.

The curtains were pulled and the room was darkened. There was a single-minded seriousness about Mrs Blackridge which made my palms clammy with sweat. Within a matter of minutes a faint voice came through the trumpet that was placed on a table to the side of the room.

'This is Jim. Are you well, my darling?'

'I am well. How do I know this is really Jim?' I asked. There were sounds of annoyance emanating from the trumpet.

'Why are you so suspicious?' The voice became louder, then continued, 'All right. I will tell you something about yourself. You are a naughty girl. You put your teacups in your cupboard.'

This made me laugh. It was a nasty habit I had acquired recently. After a rushed breakfast, I would pile up the dishes inside my cupboard on an empty shelf, with the intention of washing up in the evening upon my return from work.

While I was ruminating, I felt a sudden chill over my

head and turned to look up. The voice from the trumpet came out excited. 'Did you feel that?'

'Yes!' I answered.

'I ruffled your hair.'

I was at a loss for words. Jim spoke again. 'Are we going or not?'

I had been planning a visit to New Zealand; I had not mentioned this to anybody.

'You ought to know the answer to that,' I said. 'I thought you were supposed to see farther ahead than we can.'

The voice laughed and said, 'You may think this is far-fetched, but the spirit of St Francis of Assisi is with you. He guides you. You are under his banner. You will have the powers of healing, healing for animals. Animals in need of help will always be brought to you.'

I asked him about my mother who had passed away some years earlier. He said she was happy with her three sons and my father. I had only one brother left, the rest of my family was where Jim was.

'She sends her love to her darling child.'

'Why doesn't she come to me?' I asked.

'You dismissed her when she tried. She says she doesn't have the strength to make contact with someone who resists,' he chuckled. She was very amused at your attempt to repair garments the other day!'

'I haven't been repairing any.' But then suddenly the penny dropped: the day before I had taken out a slip only to find that the lace at the top was badly torn. I tried to mend it, but ended up piecing it together with safety-pins instead!

As I had this conversation the thought crossed my mind: Yes, this is Jim. He is all right, wherever he is, and some day I will be with him.

Immediately, the voice said, 'Finally, darling?' I felt tears rush to my eyes, and I whispered back, 'Yes, Jim. Finally.' At last I was admitting to myself that my Jim lived on, perhaps in another form, in another dimension, but the essence of Jim was still there.

'I will go now.'

'Are you saying good-bye, Jim?'

'Never good-bye. Only au revoir. Just that, my darling girl.'

That was the last time I had contact with Jim. I no longer seemed to need it. There was now an inner tranquillity. My years of searching had come to an end. I know that Jim and my family are always close to me, no matter where I am. Jim did his best, through every possible channel, to prove to me that life and love continue beyond what we call death.

Many days later, feeling light-hearted, I accepted a dinner invitation at a friend's home. After dinner we were sitting in the drawing room. I was on the sofa, with a guitar, singing one of my own compositions.

> I've often expected and been disappointed
> But now I'm as certain as mortal may be,
> That just round the corner, the very next corner,
> Something most wonderful is waiting for me.

Suddenly my friend, Evelyn, sat bolt upright in her chair and said that an air force officer was sitting next to me, laughing and beating time.

Later that evening when I returned home, I found these words scribbled on the book-marker that was tucked between the pages of the book I was reading. 'Think of me as being always with you. Talk to me as you used to do. Let us laugh at the same jokes together. But most of all, remember—that I'm waiting for you. Just round the corner . . .'

Shortly after, I began my planned voyage to New Zealand. On the way I had thought of breaking journey in India—a nostalgia visit. And here something most wonderful was waiting for me . . .

Five

COMING HOME

I can never be perfectly certain which of them was really responsible for changing my mind for me—the dog or the horse.

I had certainly never had any intention of staying on in India when I arrived in 1958, as I had only come out on a six months' visit on my way to New Zealand. I had lived in India as a child; my father, grandfathers on both sides and practically every male member of my family having been in the Gurkha Regiments. This was to be a purely nostalgic visit, after which New Zealand, where I intended settling, was to be my final destination.

After spending a holiday with my cousin, Colonel Andrew Mercer, who, after retiring from the 7th Gurkhas, had made his home in Darjeeling, I came down to Delhi where I had stayed for a spell as a child. I was putting up at the YWCA, but like most visitors to India I was quickly taken up by a number of kindly Indian families, all of them eager to show me around.

I shall not easily forget the day I was taken by a young man to visit his home. It took some time to get there since we were doing a bit of sightseeing. When we arrived he asked me what I thought of New Delhi: was it not a fine city and improved beyond all recognition since my childhood days?

I replied that it was indeed a fine city, but that I should have enjoyed it more had it not made me so sad. Sad? My friend looked at me amazed.

'What saddened you?'

'Well, for one, there was that terribly over-laden *tonga* with the very emaciated horse which we passed when we were just starting,' I said, 'and then that poor, thin pie dog with the broken leg, limping along just in front of the President's residence. Then there were those wretched performing monkeys—not to forget the bear dancing outside the Imperial Hotel.' I stopped to take a breath. The visions were flashing past my eyes faster than I could translate them into words. 'Also the lame bullock pulling a cart on the way to the Qutub Minar. And going up the hill, on the way to Karol Bagh, that poor little mongoose tied to a string, languishing in the sun. The buffalo working on the sugar cane mill, going round and round with its eyes bandaged. But worst of all, the horse. The one standing on the side of the road, with the flock of crows pecking at its back. The one with blood streaming from the sockets of its eyes: did you see it?'

'Stop! Stop!' protested my Indian friend. 'I didn't notice any of these things! I didn't see anything at all.'

He spoke the truth. He had not noticed. But then, nor had any of the milling pedestrians. Nor the ladies passing by in their chauffeur-driven cars. Not even the children playing cricket on the footpaths. They simply did not notice.

As I got back to my lodgings in an obviously worked up state, I nearly collided with the American lady with whom I was sharing a room.

'Say! What's the matter with you?' she enquired kindly. 'You look all het up.'

I told her as briefly as possible of all the traumatic sights I had seen on the way. She responded sympathetically.

'Sure. I know what you must feel like. India's a terrible country for animals. A friend of mine who came here last year said she cried herself to sleep nearly every night on account of the poor animals, because, of course, nothing can be done about it!'

I mulled over her two statements. The first was a judgement on India which I was not willing to share. I understood that coming from an environment of affluence where human hardship is kept far from the mainstream of middle-class life, my American room-mate was perhaps overwhelmed by the sights and sounds around her. It is easy to whip up one-line statements about countries and cultures and brand them forever as 'good' or 'bad'. The second statement—'nothing can be done about it'—affected me deeply. Of course something could be done. There had to be something.

I tried to unravel why there was so much indifference towards animal and human suffering in this country of pacifism and compassion. What was the explanation? Suddenly a possible answer flashed into my mind. I recalled something my mother had told me. In Switzerland once, staying in a cozy hotel close to a waterfall, she had been sure that the sound would keep her awake all night. Three nights later, however, she had become so accustomed to the sound that she found herself asking a friend whether the waterfall had dried up! It was the same with the people of India: they had grown so accustomed to the misery around them that they were immune to the distressing scenes that horrified me. The only thing to do was to shake people out of their slumber of insensitivity.

Then and there I made a vow to myself: to try to open the eyes and hearts of the people of this country—the country of my childhood, the captivating country of my memories—to the suffering around them. The vow meant that I had to remain in India. To be sure, I had a wealth of memories of India: of childhood friends, the sons and daughters of cooks and bearers from whom I had picked up Hindustani; of holidays in the enchanting hills of the North; of lazy, hot summer afternoons drinking *nimbu pani* and eating cucumber sandwiches. But these were the memories of another India—the India of the Raj—a magnificent colony of the empire. It was now a decade and more since Lord Mountbatten had handed over the reins of the country to Pandit Nehru. The Union Jack had been lowered for the last time at sunset on 14 August 1947.

Affectionate memories notwithstanding, could I actually live in a land that was not my own?

How well would I be accepted; how easily could I adapt to a totally alien culture? How long would my resolve last before I threw in the towel and decided to go back home? These were questions that I could not answer.

That night I had a silent communication with my dear, long-lost friend Jim. I asked him to help me find a solution. Towards the early hours of the morning I had a dream: I saw a dry landscape and a frail man walking with a band of dogs scampering at his side. He looked toward me and then he, the dogs, and the landscape all vanished. There followed a sensation of water, of fluidity and motion. I woke with Jim's voice ringing in my ears: The spirit of St Francis guides you ... move on, take the next step ...

So, the decision was made. India was to be the country where my future would play itself out. I was prepared for change, and I was prepared *to* change.

To begin with, I was in a quandary: I did not know where to start. Perhaps, I told myself, the best thing to do would be to seek out an animal welfare organization. I went to the local SPCA (Society for Prevention of Cruelty to Animals) and told them that I should like to work for animals. Could they use me in any way, I enquired. They put me on their committee which took up a whole half hour of my time once a month!

This was a joke as far as I was concerned. I wanted to do a great deal more. It was at this point that I was introduced to an Indian gentleman who was already running a magazine of his own. I felt having a 'mouth-piece' which could help me reach out to a wider range of people might prove very useful. 'Why should we not combine forces?' he suggested. I would save a lot of money if I did so, he advised. (To this day, however, I do not know what money I saved ...)

With complete trust I yielded to his persuasions. I put

all my energy into his magazine, he assured me that his co-editor in England would be delighted. Simultaneously, I met a Dutch doctor who shared my concerns regarding the fate of India's animals. I introduced my two new friends to each other and between us we agreed to start a new organization which we called The Animals' Friend.

TAF was inaugurated on 4 October 1959, the World Day for Animals. Its name was added to the name of the magazine already being published by my Indian colleague.

At this point I was quite unprepared for the next development. A letter arrived from England from the co-editor of the magazine with which The Animals' Friend had amalgamated. The letter showed that the writer was anything but pleased with me. I was accused of stealing her members, not to mention their subscriptions, and a number of other unpleasant things. Despite my explanations and apologies, these letters continued to rain upon me thick and fast, until I eventually decided that the only recourse open to me was to sever myself from the magazine and start a new one of my own.

Meanwhile, the Dutch doctor had found that upon the expiration of his permit to stay in India he was unable to renew it, which sent him quite unexpectedly back to Holland. Never to return.

His parting words were not encouraging: 'Well. The Animals' Friend is finished!'

'Oh, no!' I replied, looking him straight in the eye. 'The Animals' Friend is only just beginning.'

The third member of our erstwhile group was also on his way out and I was left to struggle on as best as I could without my two previous partners.

The only expense in those early days was the actual publication of the magazine. Even for this I found it difficult to raise the necessary funds (our membership was still too small to help much). I was constantly having to dig into my own pocket. Another sizeable expenditure was on the many strays that I perpetually kept finding here and there: they needed a home and since I didn't have appropriate lodgings I would end up carting them off to the SPCA which charged for each animal's bed and board.

In 1962, after several moves, I had found myself a room in a house in Nizamuddin, one of New Delhi's most pleasant suburbs. By this time I had already bought myself an autorickshaw and had designed an outsized wire-basket to hang on the back. This was my mini ambulance service for the birds and puppies of New Delhi that were otherwise left to die.

The lady whose house I was living in was not a lover of dogs in the main. On the other hand, I sneakily thought to myself, I was living in a large *barsaati* room which led onto a spacious, flat roof, and I was set apart from the rest of her house and family. The temptation, I confess, was too great and I started smuggling in puppies on a regular basis!

In order to avoid being seen I would wait a little way down the road and then give a whistle through my fingers and teeth, much like the street-lads of England. This signal was immediately recognized by my bearer, Prem, who would then emerge upon the roof carrying a small suitcase. After exchanging signals he would meet me at the appointed place and the puppy would be bundled in together with some tit-bit. Meanwhile Prem and I would keep our fingers crossed that the puppy remained as soundless as possible on its journey up the steps to my room. In this way a large family of animals found its way up to my home. There were occasional tricky moments: chatting with my landlady downstairs I would suddenly hear the small growls of puppies at play and immediately break into a paroxysm of coughing to disguise the sounds from above!

So far, my job on the new magazine was mainly editing the text, writing the occasional humanitarian article, dealing with correspondence and frequently talking to the local schools. I wanted to do much, much more however. Two of our members, Pat Dunkley and Brigette Wahi (who later became the president of the organization) kept impressing upon me the need of having our own shelter for the animals. However, the money shortage continued unabated and a shelter seemed to me like a distant dream, perhaps never to be fulfilled.

SIX

TROUBLE ON WHEELS

At the beginning of 1963, I decided to return to England to see if I could raise any funds for the cause. Shortly before my return an aunt of mine passed away leaving me a legatee. I could not believe my good fortune—at the expense of my poor dear aunt, of course. I immediately decided to spend my legacy on a car which could be brought back to India and could be used as an ambulance for the animals instead of my pitiful little autorickshaw.

I accordingly went to England, armed with a large collection of cine pictures with which I hoped to win the sympathy of the English public for the suffering of animals. The first cruel blow which fate was to deal to me occurred quite soon after my arrival.

I had driven back to London from the west country and had arrived on a cold, wet evening, as cold and wet as only England can be in the month of May. I took out my suitcase and cine projector, and leaving the rest of my luggage in the car, went into the house of the old friend with whom I was staying.

My friend, Ida, is one of those people with whom I can

pick up the threads exactly where I last put them down. Within a few minutes we were sitting in front of the fire, drinking tea and finishing the conversation we had started five years ago. I then began watching a play on television and delayed bringing up my luggage still further. Finally, tearing myself away from the fireside, I braced myself to step out into the rain. I might have saved myself the trouble. The windows of the car were all wide open and there was no luggage to bring in!

The shock was so great that for a moment I was completely stunned. Ida came out and assured me that I must have brought the luggage in and then forgotten about it. We both tore in like lunatics and searched under every piece of furniture in every room. When the full realization hit me that the car had actually been robbed, I jumped into it and drove like a crazed beast the wrong way up a one-way street to the nearest police station.

The police were very kind. My typewriter and camera, they said, would be found the moment the thief tried to sell them. As for the precious cine films—well, they were obviously no use to anyone but myself and would doubtless be found within a few days, probably inside some wayside dustbin. I tried to let myself be lulled by these false hopes.

A big jewel theft that occurred on the same night as my 'insignificant' robbery, seemed surprisingly enough to help me. A high-up official from Scotland Yard was being interviewed about the jewel case. The television people were kind enough to feel that my robbery might make an amusing contrast. Having interviewed the Scotland Yard detective, it was now my turn. The announcer came on the tube and said:

'While we hear a great deal about these big jewel robberies involving thousands of pounds, there are other robberies, perhaps involving very little in actual cash, but causing just as much heartache. We have with us in the studio tonight Miss Crystal Rogers, who has recently arrived from India on a fund-raising tour. The photographs and slides with which she had hoped to raise the funds were stolen from her car last night. Am I correct in saying, Miss Rogers, that the subject of your films were all in connection with your fund-raising mission?'

The spot-light moved to me and I gabbled feverishly about the 'mission', trying to squeeze in as much as I could into my allotted air-time.

'You're certain to get them back,' I was assured by the people at the television studio.

'You'll probably get dozens of letters and maybe some donations too!'

My television appearance brought me in but one solitary letter from a viewer in Sussex, commiserating with me in my loss. I felt that as a televisor—or whatever the word is—I had not been a conspicuous success and that the next time I got an opportunity to be on television I would get some beautiful blonde to appear instead of me. She could sit there opening and shutting her mouth appealingly while I, crouched behind her chair, could dub the words for her.

I am sad to relate that the films were never found, though I stuck an enormous notice about them to the back of my car, and carried it to the very day of my departure. Shortly before my return, I remember driving along the Great North Road and being hailed by the voice of at rubicund lorry driver: 'Lor' love us, ducks! Ain't you found them films yet?'

I had thought that without the films I should not be able to raise a single penny, but there I had been wrong. Perhaps I *sound* better than I *look*. Whereas my television appearance had been a great disappointment to me, my one broadcast on the radio service of the BBC was a joyful surprise. I had not asked for money but cheques and postal orders kept rolling in. This was coupled with the money that had come in from a few drawing-room meetings and some coffee parties. I had soon raised a thousand pounds. At one meeting in Bournemouth, a member of the audience, who had tears in her eyes, pressed a cheque into my hands, with these quiet words: 'For your animal shelter.'

I looked at the cheque and could not believe my eyes. It was a handsome sum of two hundred pounds! I began to have fresh hopes. If, I thought to myself, I could bring tears to someone's eyes without my films, then all was not lost after all. After this eventful trip, I was looking forward to returning to India. A few weeks earlier, I had shipped the car that I had bought, and had been assured that it would reach the shores of India before I did. I deplaned in India and walked confidently into the customs office in Bombay. I expected to waltz in and waltz out as I had obtained the necessary import permit from the Government of India before leaving, with the assurance from the Ministry of Finance that I should not need to pay any duty for the car upon my return. How wrong I was!

'How soon do you expect I shall get my car?'

'As soon as you have paid full duty, madam,' the clerk replied laconically.

'But I don't have to pay duty,' I answered with the air of one to whom the troubles of ordinary mortals do not apply. 'I am, you see, bringing in the car for animal welfare work. I have got exemption.'

'Prove it,' he said briefly.

'Here!' I waved my permit at him. 'This is my permit!'

The clerk looked at me as if I was a rather dense and unintelligent child.

'That is your permit to import a car,' he explained carefully, 'it has nothing to do with exemption. Where are your exemption papers?'

I stood staring at him with my mouth hanging open. I was in a terrible jam and had to find a way out as soon as possible.

'How much would the duty be?' I asked weakly.

The clerk asked the price of the car and proceeded to make a rapid calculation on the margin of his paper. He looked up shortly and said in the most indifferent voice, 'About fourteen thousand rupees.'

'Fourteen thousand!' I repeated stupidly. I could not even raise fourteen hundred at that juncture.

The man looked at me without any interest and shifted his *paan* from one cheek to the other.

'You must pay full duty or produce your exemption papers within the next three days,' he said mechanically,

Colonel G.W. Rogers, DSO, Indian Army, father of Crystal Rogers

Crystal Rogers's mother

Crystal aged five

Crystal, aged eight, and 'Roy', at Ooty (March 1914)

Crystal, October 1934

Crystal as ambulance volunteer during the Second World War

Jim Currie of the Canadian Air Force: 'the man who kept his promise'

The Mehrauli shelter grounds, Delhi 1960. The Qutab Minar can be seen in the background on the right

Inmates of The Animals' Friend

'otherwise your car will be taken from you and sold by auction.'

To say it was a shock is putting it mildly. Not only was my car not mine at this point but I had also been made to feel as small and as crushed as an earthworm. My precious little car, bought out of my aunt's legacy—and I might not ever have the pleasure of using it as my handy ambulance ...

My fried, Mr Mehra, who was the Inspector General of Police in Jammu and Kashmir, had written to a police official to give me every assistance. The official met me upon my arrival and did whatever he could think of to help me, but there was little he could do. For the next few hours I rushed around Bombay trying to enlist the help of anybody who was able to offer it. Mr J.N. Mankar of the Humanitarian League of Bombay quickly rose to the occasion. He telegraphed to Mrs Rukmini Devi Arundale, Chairman of the Animal Welfare Board in Madras, the President of our organization and also the one responsible for acquiring the exemption in the first place. Her reply to me was brief: 'Fly to Delhi immediately.'

Mr Mankar got me my ticket to Delhi and persuaded the customs to extend the three days' time-period to ten.

It was nine-thirty in the morning; the government offices in Delhi were slowly beginning to open their portals for a day of lugubrious chores. I presented myself at the office that I had been advised to go to. The Great Man whom I had come to meet had not arrived. I was given a chair in his office by his PA. Overcome by weariness, I put my elbows on the table in front of me and placed my head in my hands. I had almost started to doze when the official I was waiting for walked in.

He was pleasant enough. He apologized for having kept me waiting; sympathized with my predicament and finally assured me that he would do everything in his power to assist me.

It was not until many years later that I heard a sequel to the above story. A friend had visited the same office and met with the same official. The name of The Animals' Friend cropped up and the officer burst into loud roars of laughter. 'How well I remember some years back when Miss Rogers visited me in this office,' he chortled. 'I came in to find her sound asleep with her skirts practically over her head! I really found it necessary to reprimand her—"Miss Rogers," I said, "This is a government office and no place for sleeping!"' This anecdote was received with loud laughter by his audience, with the exception of my friend, who apparently had a strong desire to punch the fellow on his nose.

The next few days were like the strange nightmares of a fever patient. With the kindness and courtesy of an official of the finance ministry, a letter was finally traced from the late minister of finance to Rukmini Devi Arundale in which it was clearly stated: 'When Miss Rogers brings her car to India it should be exempted from duty.' The necessary papers were handed over to me and this time I returned to Bombay travelling third class by train. I arrived on the ninth day of the ten-day time period that was allotted to me by the customs office.

Still day after day passed; I ran from this office to that office, filling in forms, putting my signature to one thing after another. People told me horror stories about cars that had been left on the docks too long and when claimed had been found to be minus almost everything but the steering wheel.

I had not been feeling well since the day of my return to Bombay. There was an angry rash on the lower half of my body which was also painful. I waited a week before finally deciding to visit a skin specialist, wondering all the while what dreadful disease I might have contracted.

The doctor viewed me with horror.

'How long have you been going around like this?'

'About a week,' I said apologetically.

He ordered me to admit myself into a hospital where I was to remain flat on my back for at least a fortnight until the disease ebbed out. The condition was diagnosed as shingles—common enough, but very painful. I demurred, saying that I would be better off in my hostel accommodation, where Prem, my bearer, who had come down with me from Delhi, could attend to me better than any nurse. So that was that. I was imprisoned in my room and Prem was as good as gold. If not for him I might very well have starved. The hostel administrators persistently ignored the fact that I was a vegetarian and with utmost regularity would ply me with a diet of sausages and ham!

Meanwhile, I had no idea what was happening with my car and was growing more and more frustrated with each passing day. Finally, in desperation, I wrote a letter to a young English lawyer who I had met during my recent stay in England. He was good enough to put me in touch with one of his seniors, Ron Aitkin, who, as luck would have it, was residing in Bombay.

During this time Ron and Pam Aitkin became my guardian angels. Pam, on her visit to my hostel room decided there and then that my accommodation was insufferable and I was whisked off to their charming flat in Colaba, overlooking the sea. I had come to them a complete stranger but nothing was too good for me. I was nursed devotedly by Pam Aitkin and I had only to express a wish for anything to find it at my elbow.

Ron Aitkin sent me his own personal secretary and stenographer who took care of all the correspondence that I could not manage on my own. Ron and Pam did everything to make my visit as pleasant as possible, despite the almost constant pain I was in.

Although I had arrived in early October, it was not until the first week of December that I was finally well enough to return to Delhi. It had taken all this while for my car's discharge papers to be finally processed. The day I got possession of it I was too defeated and tired to even feel the slightest flutterings of accomplishment.

I travelled back by train, while Prem and an Automobile Association chap drove the by-now infamous little car back to New Delhi. It was a most ignominious return!

Seven

LOSING PARADISE

It still took me another week to get back to normal. The illness and the endless battle with the bureaucracy for the car had taken a deep toll on my mind. I now had only one burning obsession: to get some land to build a shelter on.

It was on Christmas day, 1963, that I was taken by a friend of mine, Major Ramachandra, an official in Bharat Sevasram Sangha (one of the leading welfare groups in India) to a Christmas service in one of the camps that he was running. The camp was in Mehrauli, to the extreme south of New Delhi. The place was more like an ashram and was being supervised by a 'swami'.

'The "swami" is actually a Buddhist monk from Cambodia,' Major Ramachandra informed me. 'He also happens to be a good friend of mine. I will definitely introduce you to him. Who knows, if he gets interested in your cause he just might offer you land right here for your shelter.'

My wildest dreams were coming true! Lady luck must have been pleased with me that day. The swami, after reading all the literature on The Animals' Friend, called me back after the Christmas service and announced that he would be happy to share his land with us!

'If you would like to have this land, I shall have to charge you a nominal rent of one rupee a month.'

My heart turned over with an almost audible bump. I was so excited I very nearly kissed him, but caught myself just in time. Middle-aged, English animal-activist kissing Cambodian, Buddhist monk turned 'swami' would have been too bizarre. I calmed myself a bit and asked to see the land.

The swami led me across the thickly cultivated ashram grounds. On the extreme north side a rough footpath led up a steep incline to a small plateau of less than half an acre. It was surrounded by crumbling walls and dominated by the ruin of an old Mughal tomb in the centre. There were no trees—only a wilderness of rocks and thorn bushes—but there was a tremendous romance of antiquity about the place. I was not prepared to notice any of its shortcomings. For me it was the site of our first ever animal shelter and as such the most wonderful place on earth!

The swami had offered me accommodation in the main building which was next to his youth hostel. My quarters consisted of one main living area and several alcoves and lean-tos which could be variously used as kitchen, pantry and even a sleeping area for Prem, my bearer. We just had to be creative about how to best use all that jumbled space.

I moved on 1 January 1964. My heart thumped wildly as I set a battalion of labourers to work on the land. So, this was it! Our animal shelter where we could fulfil our mission of serving the lame, neglected, abandoned and unloved creatures of God.

The levelling of the land was a difficult task. The rocks were so deep-rooted that they seemed unprepared to yield to anything less than blasting. I refused to be downcast and continued to view the land through a happy, rosy haze.

Soon after, a 'land warming' was hosted by us. The committee of The Animals' Friend came and rhapsodized over the surroundings. Mehrauli reeked of antiquity. Many of the ruins dated as far back as the third and fourth century BC. There was no denying the charm of the geography and I felt convinced that the animals and I had found our happy home.

A carpenter was engaged and was soon busy putting up kennels and stables. I was very, very green and trusting. The carpenter was very, very smooth and suave. I was being charged for the most expensive timber while the cheapest and worst quality was what I actually got. What I didn't know then but learnt later was that he had only recently come out of prison—where, from later reports, he shortly returned!

I went out in my 'mini ambulance' daily to bring back a new batch of injured puppies, scabies-ridden dogs, sometimes an injured cat. We would often get information from the local villagers about a hurt donkey or a sick cow. The bigger animals would have to be fetched on foot. It was not long before I had to engage not one man but two to look after all the animals that we had collected. And even two were not enough on occasion! I recall an incident where even a battery of men could not have accomplished the task at hand.

There was this bullock and a huge fellow at that. However, he was old and his working days were obviously over. Nobody wanted him, so he had wandered down the roads foraging on whatever grass he could find. When I finally saw him, I tried to piece together—from the evidence his physical condition provided—the suffering of his past days. It was clear to me that he had been the victim of an accident. Probably a speeding truck on an unlit road . . . The bullock had several injuries on his back and forelegs but no broken bones. He must have picked himself up eventually and continued in his search for fodder, which finally brought him to the open gate of a spacious house with inviting green lawns. It was after dark and nobody had seen him enter. The grounds were extensive and by the time he reached the farthest lawn he could walk no further. He had reached the end of his endurance and collapsing on the long, uncut grass, he lapsed into semi-consciousness.

His wounds were all infested with maggots and it was evident that the animal's body was giving up on him though he was still painfully alive. When it was obvious that the bullock was incapable of moving and would most certainly die right there on those grounds, we received a call from the residents of the house. Our ambulance arrived there, but it was clear that an army of twenty men would be required to load the poor beast on to the ambulance. Perhaps even that would not be enough—maybe a crane would have to be arranged. And how on earth would we get our ambulance to move with an animal that size in its rear seat? How would we squeeze a dying, semi-conscious animal into that little ambulance? It was an impossible situation. There was only one obvious answer: to put an end to the poor creature's misery by administering a painless injection. When I suggested this to the master of the house, he was indignant.

'What do you mean? You are going to poison it on my lawns? Nothing doing! I will not be party to this; we will never permit such a thing.'

'But the bullock is dying as it is. You will be ensuring an easier demise for the poor , wretched creature!' I tried to reason.

'Nothing doing, madam! You pull it out on to the road and then I don't care what you do with it, but it will not be killed on my premises!'

'But don't you see, you are causing it far greater misery by insisting on it being dragged out?'

'Misery or no misery, madam. The bullock will not be killed on my premises and that is my final word on this matter.'

However, the bullock saved everyone any further debate. While this altercation was taking place the animal had decided that its life was done and gave in to the peace and eternity of death. The master of the house and I stared at each other for a long moment. Finally I said, 'You have plenty to learn about kindness, sir. I am thankful that the bullock realized the plan you were engineering for him and decided that he wanted none of it. Good-bye!'

On the trip back to the shelter, I tried to understand the logic that had been guiding the master of the house: *Kill the beast two inches away from my main gate, I don't care, but do not kill it on my grounds.* It amused and appalled me at once.

With the beginning of the hot weather, the disadvantages of our shelter land began to become apparent. There was no water source other than a well and this was at the extreme opposite end of the ashram grounds. This meant that every drop of water for the animals in our shelter had to be brought by bucket from more than two hundred yards away, and then had to be lugged up a steep ascent. There was no electricity in the ashram which, of course, meant no fans or electric lights. As the weather proceeded to become hotter, I realized for the first time what Delhi was really like in summer.

My new home had many drawbacks but its single, greatest strength was that there was never a dull moment. Since the ashram's youth hostel was situated cheek by jowl with my own room, I never lacked company. There was always a steady stream of visitors, mostly foreign, who would come by for an overnight visit and sometimes an extended one. Some on a quest of 'finding peace'—an awfully fashionable quest in those days, and some who had just run out of money on their vacation and needed a place to rest their weary limbs.

One of these was a young man by the name of Pierre. Pierre was Swiss. I remember the evening he wandered into the ashram looking as lost as an antelope that had strolled into Piccadilly. He perched himself on a boulder and held his head in his hands. I approached him a trifle carefully; by now I was getting used to the idea that there actually are a number of unpredictable people in the world!

I asked him gently what the matter was. He replied

with more than a tinge of sorrow that he had been robbed of all his money and that he would need to stay in the ashram until a fresh dispatch of cash arrived from home, if that was all right with the swami and whoever else mattered …

Of course, I knew that his basic needs would be met by the ashram, but I also knew well that the swami was having quite a difficult time with the numbers of homeless, ostensibly 'stranded' travellers who invariably found it so convenient to 'crash out' in the ashram.

An idea struck me. One of my boys was just going on leave to the hills and I now had more animals than could be dealt with by one man alone. I put the proposition to him: how would he like to help out with the animals in return for his meals? Pierre jumped at the proposal and it seemed to me it was a good solution for all concerned.

To say that Pierre was an unusual young man would be a tremendous understatement. There seemed to be nothing he had not done and nowhere he had not been. It was quite some days, however, before the subject of flying saucers came into the conversation.

I myself had never spent much time agonizing over the possibility of flying saucers, but for Pierre, I realized, they were his raison d'être. He showed me some photographs which, he said, he had taken himself over the years. Photographs that looked like milky streaks in a dark sky; some like hazy whirlpools against the backdrop of what could have been a night sky. But, for Pierre, they were quite obviously flying saucers in the sky, flying saucers landing, flying saucers taking off and so on! He said he had hundreds of photographs of various unidentified flying objects.

'I was turned out of Israel on account of my photographs,' he told me. 'The Secret Police seized them and these are the few which I managed to hide.' He then proceeded to tell me how he had flown from Switzerland to Israel by—of all the eye-popping things in the world—a flying saucer!

'But, you see, Mees Rogers,' Pierre said sadly in his charming Swiss accent, 'I can not tell anyone about the flying zauzers because they would surely consider me mad ...'

While Pierre's money from home continued not to arrive, his own character took on more and more mystery every day. His photos of the flying objects could be considered by some as quite convincing; some of these photos he got developed by a local photographer. One morning, events seemed to suddenly become more serious.

Pierre arrived at breakfast in a frightful state of agitation, insisting that he had been robbed. Not again, I muttered under my breath!

In his hand he held his photo album, from which two or three photographs had been neatly cut. According to his own account, he had been up in the shelter with the dogs and had come down to find his suitcase open and in great disorder, and the photos in question missing. From that morning onwards he began to show signs of nervousness and kept hinting that his life was in danger.

One day Pierre returned from New Delhi proper, this time insisting that he had been followed and that someone had cut open his pocket with a razor. Fortunately, he had lost nothing, presumably since he had nothing to lose. The next event, however, took a more sinister turn.

Pierre had been sitting up in the shelter with the other boys, no doubt telling them some fantastic story which none of them could understand, when suddenly two shots whizzed over the wall in quick succession, missing Pierre by inches and going through a tin of kerosene standing nearby. The boys leapt up and ran to the boundary wall, only to see a car disappearing down our narrow lane.

Naturally, I rang up the police. They had already heard about Pierre's missing photographs and possibly about his slashed pocket too. 'Yes, yes, we will certainly come and investigate,' they assured me.

They never did, as a matter of fact, so perhaps they didn't believe the story to be true. There was nothing imaginary about those two bullet holes, though. They were as real as daylight. Meanwhile Pierre, who was becoming even more bizarre than before, had acquired a monkey who came to all meals and ate almost as much as his master. Pierre's money from home was still to arrive. He was beginning to feel the heat and his work output was quite miserable. I began to get quite restive and even his 'flying zauzer' appeal failed to make me want to keep him any longer.

Whatever was the truth of Pierre's past went out the door along with him some days after the shooting. He was the topic of animated conversation for a few days: Who were the people in the car who had tried to shoot him? What was the reason for it? What kind of activity was

Pierre truly involved in? And then, like all sensational stories, he passed out of our lives. Just as well, because by now other difficulties had begun to enmesh me and I was thankful that I didn't have to bother my head about Pierre's mental health and physical safety any longer.

I had almost forgotten to notice what it was like living without water, without electricity, without drains. I suppose some inner, hidden rationale was at work: if my precious animals could live without all these luxuries, why couldn't I? Anyhow, in addition to the discomforts, the clouds of hostility were gathering over my head. It was a development that took me quite by surprise. Once again I was to learn, the hard way, that it takes a lifetime to understand the nuances of different cultures.

The local Muslim population of the area, mostly engaged in farming, had suddenly discovered that an English 'memsahib' was keeping animals—and worst of all, dogs—in a very holy place where there had once been a mosque and where there was still an old Mughal tomb.

They sent their representative to see me. I was out. In those days, as I was my own ambulance driver, I was more often out than in. The representative was seen by the swami instead and the former stated his case.

'The English lady is desecrating a sacred place by keeping animals there, and she must be asked to discontinue her activities at once. Or better still, ask her to leave.'

The swami was very diplomatic. He regretted his inability to ask me to go since he had committed himself for a period of three years and could not go back on his word.

The young man left greatly dissatisfied, saying he would call again. In due course he did—with exactly the same result. I was out once again. Finally after a lapse of some months, he came by a third time with the same proposition. He was again interviewed by the swami. This time the message I received carried a veiled warning . . . they would take recourse to the police as a means of evicting me, they mentioned. But both the swami and I knew that the police could do nothing as I was not doing anything to break the law. This time, however, I felt the hidden menace behind the message. I little knew then what lay ahead.

If I believed in curses, and I am not quite sure whether I do or not, I should have said that from this moment onwards we appeared to be under a curse. The walls of our shelter started to collapse and just as we repaired them, they collapsed again; the cages and kennels, poorly constructed out of cheap timber, kept falling to pieces; our barbed wire was uprooted and the gates fell off their hinges. The shelter witnessed pandemonium. With the breakdown of the cordons that kept the animals in their sanctums, we found that it had become impossible to herd them and they were running wild all over the area. The water issue was becoming a monumental problem. A succession of three well-diggers were asked to dig us a well, but although we wasted a lot of money over the project, each one ran off without finding us any water.

Strange things happen when one leads an 'off-beat' life, but this was getting to be beyond the realm of strange.

The animals—particularly the dogs—started to die without any apparent reason. No unexplained virus, no contagious disease that they could pass to one another, no reason at all. A puppy that had been frolicking the night before would be found cold dead in the morning. We ruled out poisoning. Anyone who could get in to do so would have caused a huge commotion amongst the dogs and would have certainly been seen.

One moonlit night, the boys were sitting around in the open compound chatting about the day's doings when they heard a sound like a heavy chain being rattled, coming from the direction of the well. They rushed to the scene thinking that a dog might have possibly fallen into the well and was rattling the chain in an attempt to get out. Although waterless it was still very dark inside the well. One of the boys picked up a long pole and gently probed the bottom to see if he could feel anything. At this point, to quote their united testimony, 'the pole was seized by something invisible and was flung into the air; simultaneously a large part of the well's upper wall collapsed'.

The boys were scared out of their wits and fled to their huts. They were followed by our throng of street smart stray dogs (normally fearless and proud) with their tails between their legs, whimpering like cowards. Soon I began hearing stories from the boys: 'Miss-sahib, we hear voices at night. When we go out to investigate, there is no one there.'

'Miss-sahib! Miss-sahib! Last night there was a shower of stones on our huts. When I went to see what was happening I found no one there, but still the stones were falling from the sky!'

By now, we could even boast of a ghost: a white figure which floated round the premises in the moonlight . . . Frequently, I would hear the shrill cry of one of the boys piercing the quiet of the night—*'Bhoot hai! Hai Ram! Bhoot!'* and this would be followed by the hushed yet frantic recital of the Hanuman Chalisa to calm them down and ward off the evil spirit.

A Dutch lady calling herself a Christian mystic was visiting the ashram at the time. She also slept inside our shelter grounds. She remained there all night in a posture of meditation during which she was presumably going through a series of remarkable revelations. I don't know whether she actually saw the ghost. Her presence was, however, unnerving to say the least, as she kept warning me in tones of intense entreaty: 'Go! Go! Quickly before it is too late!'

Having nowhere to go I was unable to follow her advice. 'It is evil. Evil, I tell you,' she said sneaking upon me late one evening. 'Can't you see how all the animals are dying? Soon you will not have a single animal left in this shelter. What will you do then?' On another night when the dogs were barking and howling, she exclaimed, 'Oh, the poor things. Can't you hear them pleading to Heaven for mercy?'

I hesitated to suggest that it was the charming new bitch which had just been introduced to the kennel who was having this effect upon the unrequited passions of the male dogs. It would have seemed indecent and almost blasphemous to shake her faith in the belief that there was some form of evil amongst us.

But then one evening a very odd thing happened which I could definitely not put under the category of perfectly normal events. By this time I had engaged two brothers, Kundan and Mathura, to deal with the animals. Kundan possessed a small harmonium which he would play at the end of the day while colleagues and friends squatted around him and sang. It always made a pretty picture. Even I, relaxing in my room after a long day of labour, would listen to the gathered voices singing songs—unfamiliar to me, yet, somehow soothing.

On this particular occasion, a group of them were sitting around singing, when suddenly, Mathura, at that time a boy of seventeen or so, apparently fell into a state of trance. In a voice quite unlike his own he began intoning and calling his brethren to prayer. The rest of the boys stared at him open-mouthed. Quite unconscious of the effect he had caused, Mathura continued his intoning using the Muslim word *Khuda* for God instead of the Hindu *Bhagwan*.

By now everyone who was witnessing this phenomenon was thoroughly perplexed, not to say frightened as well. They decided that the spirit of a deceased Muslim priest or fakir, probably of the one buried in the crumbling tomb, had taken possession of Mathura's body. When the boy came to, he was in a state of high, unexplained fever and he remained like that for the next three days.

This whole business was beginning to wear me down.

It was not the so-called psychic manifestations which upset me. These I rather enjoyed though I regretted not having the pleasure of encountering the ghost myself. This was not for want of trying. To the great disapproval of the staff, I had taken my bed up to shelter in order to witness anything curious that might happen. Unfortunately nothing went bump in the night as long as I was awake, and once boredom set in, I fell into a sound sleep.

What was making me anxious was the state of the animals and the fact that despite every attention, they continued to die daily for no apparent reason. What was the use of saving animals from cruelty, neglect and starvation if they were only going to die from some cause unknown a few days later?

*

Things were just going from bad to worse. The rains set in and one day we discovered to our horror that the whole area was swarming with scorpions. A few days later we had an invasion of snakes. No less than five of the dogs and a donkey succumbed to snake bites. A young boy was also bitten, but thankfully he survived. I doubt if my conscience would have been able to cope with the loss of a human life as well.

Our dear old donkey, Danny, fell into the waterless well and was so badly injured that he had to be put to sleep. An epidemic of distemper carried off all of our most attractive puppies and there was a growing atmosphere of friction and hostility amongst the staff. I began to accept that there was something very wrong with our once happy and productive shelter but, I must say, I had no solution to combat the 'forces' at work. I understood that I would ultimately have to bow to the wishes of the local Muslim farmers. I would have to respect their sentiments and find other options but I was still not ready to make the move out of the shelter. It was Jill who somehow clinched the matter. Jill, an English girl who had been working with us for some time in the shelter, had turned out to be a most precious asset for me. There was nothing she could not do: be it giving an excited bull an injection or putting up a gate, everything seemed within her ability.

One day she brought up the subject, tentatively at first, but then with a little more conviction.

'Crystal, I do wish you could get away from this place.'

'On account of all our present misfortune, you mean?' I shrugged my shoulders and tried to dismiss the entire topic. 'Presumably this will all pass and soon enough we can get back to the way we used to be.'

'No, I don't think it will be that simple.'

Something in her tone arrested me. 'What do you mean?' I asked.

'There's something badly wrong here,' she said in a voice full of firmness. 'There's something really evil that you are up against.'

'But I thought you really loved it here. It's beautiful and historic and everything that we could have ever wanted it to be.'

She spoke again with great emphasis as I stared at her. 'I wouldn't sleep in that shelter of yours for all the money on earth! I can only hope that I and the rest of us are merely imagining these things, but I do believe that is not true.'

It was around this time that we managed to get some publicity in the press. I had gone to them in desperation, wondering if anyone reading of our troubles might be induced to give us some land. The press was not particularly interested until I came to mention our 'ghost'.

After that they were all attention—apparently ghosts are popular with the press. Perhaps they never got to know them intimately enough. Anyway, one of the papers gave us front-page coverage with a picture of 'the haunted ruin where the ghost is said to walk'.

We had quite a lot of visitors after that. Mostly people interested in ghosts more than animals. Two wandering swamis turned up and after performing some reportedly sanctifying rituals assured me that I would have no further troubles. Some English friends brought in an Anglo-Catholic priest who read the Christian rites of exorcism and sprinkled holy water around the shelter.

None of the above seemed to have any effect. The manifestations continued unabated. Perhaps our ghost was not very religiously inclined. I could only continue to hope that one day we would be able to find some alternative land.

*

One often thinks afterwards that some events were planned beforehand; so easily do they all fall into place. In this case the events were all set in motion by a mad dog—a small dachshund who had had the misfortune to go rabid in a village beyond Azadpur, which is the other extreme of Delhi. It was late evening when I got the telephone call. The owners of the dog sounded panic-stricken and I make it a point of principle to never refuse a call for help.

The village was over twenty kilometres away from Mehrauli and it was already after nine at night when I set out. I was never to forget that night. The moon was at its full, and the road was quite deserted. As I went past Kingsway Camp, on my way to Azadpur, my attention was caught by an area of wooded land to the left of the road. The trees were large and looked very old. Everything looked pristine and natural, and so far untouched by the hand of man. What a contrast to the rocky, treeless land around Mehrauli. And what a wonderful spot for a shelter, I thought!

Soon after, having reached the village where the mad dog had to be picked up, I dismissed all other thoughts from my head. Having got the poor little victim into the necessary cage, I drove off and dropped him off at the SPCA, where his owners wanted him kept under observation.

It was two days later that I found myself stuck in a traffic jam on Minto Road. A small, stout Indian gentleman who was going in the same direction as myself, stuck his head in the window and greeted me.

'Hullo, Miss Rogers!' said the owner of the head. 'I don't know if you recognize me, but I'm a friend of the Swami. I have even seen your shelter some time back. If you are going my way, may I take a lift?'

'Of course,' I said and opened the door.

'Well,' said my passenger, as he cheerfully settled himself into the seat beside me, 'how are things at the shelter? All is well, I hope?'

'Anything but, I'm afraid.' I truly was feeling dismal. 'We seem to have some sort of voodoo, you know a curse, over us. We are simply swamped in bad luck.' I went on to outline some of our worries. I finished by saying that I would do anything to move out of there.

'Have you seen any other land that appeals to you?' my passenger asked.

I answered that we had been hunting around but everything so far seemed beyond our means. I added wistfully that, in fact, I had just recently passed by a piece of land that I would love to possess, but I was quite sure that it would not be available. He asked where it was and I described for him the location.

'Oh! There!' he showed enlightenment. 'That wooded land is a group of very old orchards that have been around, unchanged, for years. I hear the fruit is let out for auction every six months. You might be able to get one of the orchards but, I would imagine, only for temporary occupation.'

My heart skipped a beat. It didn't matter that it would be temporary . . . at least, for sometime it would make a peaceful home, until I managed to find a more permanent solution.

And so it came about that two days later I found myself heading back towards 'the promised land' with my new friend, Captain Gupta, the passenger from Minto Bridge.

We went down a straight road and ended up in the heart of the orchards. There was a sense of peace amongst those ancient, fruit bearing trees that stilled my mind. I felt the pressures of my disturbed life evaporate and disappear, over the low, shady mango trees. We walked on until we came to a wooden gate where a middle-aged man greeted us. He was already known to Captain Gupta. In a few words in Hindi, the Captain explained my situation. The man seemed at once sympathetic and inclined to help. The gardens, he told us, were to be auctioned in about a week's time. The one next door to his should be ideally suited to my needs. He took us to see it.

If Adam had suddenly led me into the Garden of Eden I should not have been more impressed!

The orchard comprised about two acres of thick woodland. Mulberry, mango and guava trees abounded. And to my utter delight there was a well in the middle of all this, with water, plenty of grass where the cattle could graze and, wonder of wonders, electricity close at hand! Was it possible that such a place could ever be ours?

Nothing ventured, nothing gained. So off I went to the offices of the land and housing commission, New Delhi, which I was told were the official authorities concerned. I was given audience by a clerk to whom I stated my case. He was sorry—those orchards were intended for fruit-growers only and animals would never be countenanced there. He must have recognized my look of plain despair: 'You can see the land and housing commissioner if you wish to, but I don't think it will help much.'

I was ready to clutch at any straw and immediately jumped at the invitation. I was led down to the ground floor, and ushered into a big room where a conference of some kind was already in progress. I started to back out apologizing for my intrusion, but was politely invited to take a seat at the large table around which everyone was sitting.

'What is your problem, madam? Is there anything I can do to help you?' I was being addressed by an elderly man in a large turban, who happened to be the commissioner.

I poured it all out in a rush. The state of the animals, the objections of the Muslim farmers, the impossibility of our remaining on our present land, the drastic situations I found myself up against. When I paused for breath, I found myself looking into a circle of pleasant but slightly bewildered faces.

The commissioner was sympathetic. 'Well, gentlemen,' he said, looking around the table for agreement, 'it seems like this lady is doing everything in her power to help the animals of our country—under very difficult conditions, I might add. I would think that it is our duty to do something to help her.'

There was a chorus of assent. An official was summoned and was told to see to it that I was handed whichever orchard I desired. 'And also ensure that she gets it at the most nominal price.'

I drove back to Mehrauli. My mind was in a daze. The Promised Land was in sight, at last.

On the day of the auction, Nalin Mahajan, one of our oldest members in terms of time but youngest in years, accompanied me. I was as nervous as a kitten. Would the commissioner keep his word? Suppose someone else bid higher than myself? I need not have worried. Oddly enough, no one seemed particularly interested in the orchard that I wanted. Soon enough the auction began.

'Five rupees! Five rupees!' A half-hearted voice cried out.

'Say fifty,' the clerk from the land and housing commission whispered in my ear.

'Fifty rupees!' I shouted back obediently.

There was dead silence. No further bid was made. The Garden of Eden was ours!

Eight

THE LOST SOULS OF DELHI

Our move was made on 8 May 1964, exactly a week after the auction. Our old shelter land and the new were extreme opposites in every way. The rocky barrenness had given way now to a luxuriant green; the fear and despair were replaced by calm and new hopes.

The SPCA assisted us in transporting the animals to their new home. With the summer fast approaching, the one thing we appreciated most upon our arrival in the new premises was the ice-cold water that we could dredge up from our own well!

We had to camp out, to begin with, but the boys started work at once to put up their own huts, while for the time being I was to reside in a tent. The work of supervising the shelter was left primarily in the capable hands of Nalin Mahajan, our young-old member, assisted from time to time by other members of our committee. I, in the meantime, decided to vanish from the scene for a few days to Kashmir where I worked a while with the Kashmir Animal Welfare. Upon my return I was to find a quaint little wooden house waiting for me: my home. It was also our dispensary and office all rolled into one. For this I shall always be grateful to Nalin.

It would not be quite true to say that everything in the 'Garden' was lovely and hunky-dory. The Delhi summer was once again oppressive, even in that cool spot, and brought with it the insect pests that are the nightmare of every animal owner. There were flies in the millions, ticks, and every other creepy-crawly known to man. The animals, regardless of all the scourges, were happy. The ponies and our two buffalo calves, Bonny and Bhaiji, now had room to stretch their legs, and the dogs ran about in perfect freedom, mingling quite peaceably with our newly-acquired cats which had recently emerged out of kitten-hood.

Asha and Pasha were both dead white, except for grey tails and ears. Asha had been brought to us as a tiny kitten soon after we arrived. She was in the middle of having a fit at the time. We didn't think she had much of a future and out of sheer sentimentality the boys named her 'Asha' which means hope in Hindi. Somewhat to our surprise, she recovered. Her owners collected her and took her back. Three days later they returned her in the middle of another fit. Asha recovered again, but this time her owner did not come back. I have always been baffled by an owner who disowns a pet when there is the slightest trace of an unusual malady. Would these people dump a sick mother, a blind father, an invalid brother for reasons of their own convenience? The answer to that I would earnestly hope is a no. (But something tells me it could just as well be a yes.)

Pasha, who looked almost like Asha's twin, oddly enough was brought from the other end of the city. These two girls were always together and the only way to distinguish one from the other was by their tails: one had rings while the other didn't. Anyhow, we kept them, feeling they were both attractive enough to warrant finding homes for them.

Two other attractive babies in our nursery were Mina and Mooni—a set of adorable baby monkeys. I had found them clinging to each other for dear life in a pet shop in front of the Jama Masjid in Old Delhi. I bought them to avert the terrible fate that awaited them, either as performing monkeys in the street or in the hands of vivisectors.

Mina and Mooni like all babies loved their bottles. In every other way they were surprisingly dissimilar. Mooni was inquisitive, bold and mischievous. Mina was gentle and timid and at the slightest provocation would pucker up her tiny face as if she was ready to cry.

Vigin, the wife of Kundan (one of our boys), was, like her husband, a true animal lover and was particularly good with all young things, so I naturally put her in charge of all the babies. The two monkeys made their instantaneous choices: Mooni decided that Vigin was her mother and Mina chose me. They clung around our necks like small fur stoles and though happy to play with each other, always came back to each of us respectively to be comforted and mothered.

Another delightful baby was our lamb, Laddu, who was truly sweet! Laddu was brought to us by a local butcher who had had a change of heart upon seeing Laddu being extricated from his mother's womb in the slaughterhouse just after she had been killed. Laddu would have met with the same fate had the butcher not decided, at that very moment, to take a diversion from his profession and instead follow his heart.

With great success Vigin brought him up on a bottle and Laddu reciprocated with love and gratitude by following Vigin just about everywhere. He must have been about four months old when a terrible thing happened. One of the people who tended the neighbouring orchards saw Laddu nibbling on some fruit that had fallen to the ground. The fellow, in a fit of unreasonable anger, picked up a brick and hit the little lamb on its head. Laddu dropped like a stone and lay still.

This happened on an afternoon when we were conducting a pet-show cum fete in New Delhi. Vigin arrived, her face swollen from weeping, to tell us what had happened. The fete had just got over and we drove back with her. I cautioned the boys that Laddu's body was to be buried before Vigin should have the chance to see it again.

It is difficult to describe our amazement when the boy who had been detailed to bury the lamb came to tell us in great excitement that the 'deceased' Laddu was still breathing! Vigin and Kundan were overjoyed. They took the lamb to bed with them and administered drops of glucose water to little Laddu all through the night. The treatment worked; just twenty-four hours after he had been struck on the head, Laddu opened his eyes, looked straight at Vigin and belted out a lusty *baa-aa*.

As time went on more and more motherless babies came to the shelter. Litters of puppies were left at our gate, wrapped up in paper bags. Small children came in clutching tiny kittens which they dumped on us. Saddest of all were the small victims of cruelty which were brought in all too often—usually too late. Day by day our family of animals increased. At the same time it became more apparent to me that if one felt compassion for animals one must naturally extend the same sympathy to humans as well. I was constantly picking up people from the side of the road and carting them off to hospitals.

I had witnessed man's inhumanity to man at a very early stage. Once, on my way to my printers, I spotted an old man lying on the median in the middle of the road, apparently fast asleep. Apart from mentally commenting that this was indeed a funny place to sleep, I did nothing further. On my way back, an hour or so later, when I noticed that he was still there, I began to feel that there might be cause for alarm. I propped my bicycle against a wall—this was before I owned a car—and walked towards the man. He was lying very still, face down on the cement. I shouted across to a tradesman sitting outside his shop on the busy pavement opposite.

'What is the matter with this old man?'

'Nothing, memsahib. The old man is okay,' he shouted back cheerfully.

Another man crossed the road and joined me. He lifted the parched, old hand and shook his head. I touched it too and found it stone cold.

'He is dead!' I said, feeling my throat go dry. 'What should we do?'

'Telephone for the flying squad,' said the man quickly,

'you'll find the number on the cover of the telephone directory.'

It was evident that he did not want to do this job himself. So I telephoned the flying squad as directed. After a lapse of some time a constable appeared who seemed to have some medical training. He announced that the man had probably been dead for around four hours.

Dead in the middle of one Delhi's busiest streets and nobody had noticed!

Then there was the case of old Likhan. I had befriended this old-timer during my early days in Delhi, when I used to sit in the Jantar Mantar gardens and read, write and pass the time of day. Likhan was a sweeper at the gardens; I would watch as Likhan diligently gathered the dry, fallen leaves. As was inevitable I had already adopted a local Jantar Mantar dog and the dog went on to adopt Likhan and so it was that Likhan and I adopted each other. It was the beginning of a firm friendship. Likhan was not only a genuine animal-lover but also quite an ardent anglophile, having served in his youth under the British.

One day, Likhan came to me in great trouble, with tears streaming down his face. His son had been run over and killed by a speeding truck. His mutilated body had been brought home, at the sight of which his son's young wife, seven months pregnant with their first child, had gone into sudden labour and had produced a premature baby. Could I possibly help him to get compensation for the young widow and the baby?

I was only too anxious to help but had little idea what to do. I went to see my friend Ramji and he suggested a free legal service for the poor from which I was allotted an advocate. The advocate, well into his seventies, did not inspire much confidence in me. His first act was to accept the accused truck driver's line of defence that he had not run over the boy; instead the boy had actually slipped to his death from the back of the truck after taking an unauthorized lift. Before I realized it, he had already got the young widow to put her thumb print on this statement.

Likhan was outraged. How, he asked, could his son have fallen off the *back* of the truck when the boy's blood and brains were found on the *front* wheel? I made inquiries from the police who had been at the scene of the accident. The police inspector agreed with Likhan's story. Yes—he himself had seen the blood and brains on the front wheel of the truck. But now the wheel had been sent off in a sealed container to some unknown destination for chemical inspection.

We waited for the report to come in. When it did, it turned out to be a terrible let down for all of us. 'No trace of human remains,' the report stated. The inspector shrugged his shoulders heavily as if to say there was no more he could do.

The case dragged on for three weary years. I accompanied Likhan and his family to court whenever I could. I remember once waiting for over two hours shooing away flies on the benches outside the Delhi Courts and wondering what the delay was all about. I enquired from the clerk of the court who looked surprised. 'Don't you know, madam,' he said, 'your advocate said he was not in the mood, so he has gone home.' I wish someone had told us that earlier!

On all these trips to the courts I would look around at the patient, weary faces of the poor, coming day after day, week after week, year after year, hoping for justice. Did justice ever visit them, I wondered. But the one lesson I learnt, which has been a very valuable one, is that patience is a remarkable quality and one of the precious treasures that the East can export to the West.

And then, quite suddenly, one day I found that we had lost the case. I had been unable to attend court that day but rang up Mr Naresh, the advocate, in the evening. I was literally shaking with rage as I realized that he must have been hand-in-glove with the enemy from the very beginning. He refused to speak with me. I had no idea what had transpired in the courts. What was I to tell Likhan's daughter-in-law when I knew so little myself?

The battle for Likhan's young daughter-in-law and grandchild which had been fought for so long, ended abruptly, leaving everyone drained, disheartened and smouldering.

*

Nazir's tale will remain with me forever. On one afternoon as I passed by the general post office, I noticed this crumpled human figure with his head practically falling into a gutter. I stopped and turned back. He was a sickly boy of about eighteen and at that moment he was unconscious but breathing. I started inquiring about him from a young man who had a cycle repair shop across the road.

'What's the matter with this boy?' I asked him.

'Memsahib, he is ill, I think.'

'How long has he been ill?'

'About four or five days.'

I asked whether anyone was looking after him; did he have a family; was anyone bringing him food and medicine?

'Memsahib,' the young man gave me a wry smile, 'who will care?'

'I will,' I said with resolve and went off to ring for a police ambulance.

I accompanied the boy in the ambulance and we went to the nearest hospital where he was taken in without any fuss, probably on account of the police being there. For a few days the boy's life hung in the balance. He was given saline drips, blood transfusions and every modern method of medicine was used to save his life. When he regained consciousness one thing was all too apparent. The boy was simple-minded. All he could tell me was that he came from Lucknow and that his name was Nazir.

I visited him regularly over the next few weeks. I was anxious about his future and voiced my concerns to a nurse at the hospital.

'No need to worry,' she said, 'as soon as he is physically fit he will be handed over to our welfare officer and will be rehabilitated.'

Finally the day came when I discovered that Nazir was no longer in the ward. I was directed to the office of the welfare officer who was to have been dealing with Nazir's case.

'Oh, yes … Nazir,' she said, thumbing through her records. 'That was the boy who was sent to me yesterday. Well … I had intended sending him off to the poor house but then I found that he had come in a police ambulance, so of course I could not send him without their permission. Then I decided to send him back to the ward.'

I rushed back to the ward and located the sister on duty. 'Excuse me, sister,' I said, 'I want to know about a boy, Nazir, who was in bed 14. I was told that he would be rehabilitated by your welfare officer, but she tells me that she had to send him back here because of police permission or something.'

She had been looking vague and puzzled all this time but her eyes cleared up at once on hearing my last remark. 'Oh, yes! That boy. But he was discharged!'

'Discharged? But why?'

'Because we needed police permission to send him to the Poor House which might have taken months. And we couldn't possibly have kept him here indefinitely.' She added mechanically, 'We haven't got a million beds.'

I felt my temper rising dangerously. I took a deep breath and started speaking very deliberately. 'You mean to tell me that you let that boy go off in the middle of winter, with no home to go to, no clothes, no money and not even the wits to beg? You just told him to go? What you have done is tantamount to turning a three-year-old child out into the street.'

My voice must have risen, for a doctor strolled over to us from the other side of the ward.

'What's all the commotion here?' he asked irritably.

'That boy, Nazir,' I said, 'the one who was in bed 14. He was discharged yesterday with no home to go to and he has the mental age of a small child!'

'Oh, that one, Nazir,' said the doctor indifferently, 'he was a lunatic, was he? I didn't know ...'

I dragged myself out of the hospital feeling very defeated. Nazir was gone and I didn't know where to find him again. Another lost soul walking the streets of Delhi.

*

It has always been a source of amusement to onlookers to see me running after a diseased, injured dog which does not belong to me. I have often been asked: 'Is that your dog, madam?' When I hasten a denial, the response I get is: 'Then why are you running after it? That is dirty, filthy dog; I can give you good breed Alsatian puppy. You want?'

In the same way it is difficult for most people to understand why I should bring sick and starving people to hospital unless they are in some way connected to me. After all, time is precious, I have been advised often, and should be spent wisely on attending to only those who are of immediate value to one.

My human patients are invariably as filthy, diseased and unwanted as my animal patients. In a hospital situation of the kind I have described earlier, I am invariably asked the stock question: What relation is the patient to you, madam? And my stock reply has become: Uncle, aunt, or cousin, depending upon the sex and age of the person! It always amazes me that the answer is accepted without further question despite the obvious differences of racial origin between me and my so-called relatives!

Nine

FIRE WITHIN, FIRE WITHOUT

If I have likened the new shelter to the Garden of Eden, I was soon to discover that it possessed more than one kind of snake. The first was of the common or garden variety, which for the most part kept out of our way. The other, which was far more deadly and dangerous, was both invisible and insidious. Its genus was jealousy.

I suppose it was hardly possible for us to be given a fruit orchard—at a fantastically low rent—without provoking jealousy from certain quarters. Worse still was the jealousy that sprang up amongst our own staff. The cause of the latter was mainly a young, south Indian boy by the name of Manikam whom I had employed shortly before leaving the previous shelter. Apart from being incredibly good-looking with a body that any artist's model might have coveted, he was also good at everything to which he put his hand.

As a worker he surpassed everyone else, both in what he could do and the speed with which he could do it. He was amazingly strong with muscles like whipcord, and I have never seen anyone who could run quite as fast as him. He could make the dogs' chapatis in half the time it took anyone else, mend all things electrical that went wrong, drive our little ambulance and even do running repairs.

It seemed that there was nothing that Manikam could not do. On several occasions when other members of the staff were away or sick he shouldered all the extra burden of responsibilities without showing any signs of fatigue.

To say that Manikam was extraordinary in every way would not be an exaggeration. During the winter he dug himself a large, underground room in which he slept with a battalion of dogs for company, emerging in the morning quite warm when everyone else was shivering. For the summer he built himself a house high up in a tree and reduced the number of his canine companions to a maximum of two.

Manikam's main desire in life seemed to be to please me personally. I had only to ask him to do the smallest thing and he would be off like the wind. His failing, I understood, was the desire to be my blue-eyed boy at the expense of the rest of the staff. This, of course, I didn't stand for and refused to give him the special preference for which he longed.

It was inevitable of course that jealousy should spring up on both sides. A campaign of tattling started. Most of this originated on Manikam's side. He begged me earnestly to take care of my money, not to leave it lying around, not to allow myself to be cheated and not to believe everything that I was told.

A rather dramatic theft of money from my room cast a gloom over the shelter and everyone eyed everyone else with suspicion.

A story then took wing that Prem, my old bearer, had sworn to kill Manikam while Manikam had in turn sworn to kill him! Each slept with a heavy stick beside him at night, apparently genuinely afraid of being murdered and arming himself for the attack which never came. The whole atmosphere at the shelter became dark and menacing; all of us waited for the climax to this bizarre development.

Prem asked for leave to go to his home in the mountains and I gave him permission; he left taking his young wife with him. After his departure yet another story surfaced in which Prem was supposed to have been stealing from me regularly ever since he had first come to me over four years ago. If I went to his home in the hills, the story ran, I would find his entire home furnished with the objects he had stolen from me.

I didn't believe a word of this since I had always found Prem fantastically honest and loyal to me. I disliked this kind of slander and wanted to put a permanent end to it. I could only do so by actually going to Prem's house and coming back with a definitive report on the situation which I would then share with the rest of the staff. However, even if I went there, a European travelling in that part of the country would be spotted from miles away and the guilty party would have plenty of time to hide his loot before I finally appeared on the scene.

After much deliberation I hit upon a plan: I made a number of visits to the *darzi* and to the bazaar, keeping my purchases all tied up in a disreputable old blanket which I hid. I then made the announcement that I was going away for a few days to stay with a friend. I collected my rather peculiar luggage and made for the railway station. For safety's sake and in case I was arrested as a suspicious character I took an Indian friend with me who was keen to take part in the adventure.

We travelled by train as far as Rampur, where we caught the bus that took us to the foot of the mountains where I accomplished my magical transformation. I asked if I could change my clothes in the empty bus which had been driven into a deserted yard; an Englishwoman got in and a Pahari woman stepped out! I wore the regulation dress of the hill peasant woman—long shirt, with a very full skirt down to the ankles and a cloth like an abbreviated sari tied around the head. I had also taken the trouble to colour my skin a fawn brown and a little bit of coloured black hair could be seen in the front, over the forehead. Added to this I wore a heavy silver necklace, earrings and bangles, not to mention tinkling anklets, while my pièce de résistance was a pair of old-fashioned, round, steel-rimmed spectacles tied together with a dirty length of string and worn slightly askew over my nose!

I was quite sure that my own mother would not have recognized me. Certainly all the local peasants that we passed on the fourteen-kilometre hike up into the mountains did not give me so much as a second glance.

I was obliged to take the pony-men into confidence, since otherwise I would have had to remain dumb the whole way up. My companion informed them that the memsahib was on her way to enact a little play, a concept they all understood and which appealed well enough to their sense of fun.

At last we reached Prem's house. There was no one on the horizon as we approached the modest hut. One of the pony-men knocked at the door. I slumped on the horse in a manner befitting a tired old lady of the mountains and my friend averted his face so as not to be recognized. Presently Prem opened the door. For about half a minute he stared at me perplexed and then let out a whoop of joy. 'Miss-sahib!' he cried, and flung open his door with delight.

Needless to say, I never discovered a single pin of my property in Prem's house. My friend returned to Delhi the next day but Prem persuaded me to stay and I was treated like visiting royalty for the duration of my stay. When I returned to Delhi it was with Prem and with a complete sense of confidence to counter all the mud-slinging against him.

*

It was shortly after my return from the hills that back in the shelter, one night at about 11 p.m., as I sat writing at my desk, three men entered my room. I looked up from my papers and stared into the faces of men who looked like they were on the police 'Wanted' list. Keeping my wits about me, I enquired their business.

'We were just passing . . . saw your light on and wondered what was happening in here,' replied one of them airily. He obviously thought that was quite a reasonable explanation and that I should be satisfied with it. 'What is this place?' he added by way of an afterthought.

'It's a shelter for stray and starving animals.' I was hoping that one of the dogs asleep in my room would come out and growl at him.

'Well, then, could we see the animals?' asked the talkative one. Apparently he saw nothing peculiar in being shown around at eleven o'clock at night.

'I'm afraid it's much too late for that,' I said with some asperity and didn't much care for his next question or the way he asked it.

'Are you alone here?'

'Alone? Certainly not,' I lied cheerily. 'There are workers all around. There are two of them sleeping right now inside my kitchen,' I said pointing towards the right. 'I can't show you the animals at this time, but I can show you my workers. Would you like to see them?'

I hoped he wouldn't call my bluff. The staff and their families were there, of course, but they were all in the quarters and these were too far off: the boys wouldn't have been able to hear me even if I shouted myself hoarse.

My three visitors had had enough. They trooped out promising to come back again to 'see the animals'. I saw them make their way to a gap in the hedge where they were joined by two other men. They all talked for a while and then hastily disappeared.

A few minutes later one of the boys came racing up from the staff quarters. Someone had heaved a brick onto the top of his hut and he wanted to know if I had seen anyone. I told him about the men and we both agreed that there might be some dirty work afoot and I had better call the police.

I did so, though I didn't feel that at this stage the police could be of much help. Half an hour later a sleepy policeman arrived but by now, no doubt, the disturbers of our peace were far away.

It was not many days after this that we had our first fire. I was woken at about 2 a.m. to hear the baby monkeys chattering frantically. I looked out my window and to my horror saw a sheet of flame rising from the staff huts. I rushed to the phone and dialled the fire brigade.

A sleepy voice enquired as to the location of the fire.

'Pambari Road,' I said, 'opposite the new Police Lines!'

'What road?'

'Pambari.'

'Spell it,' said the voice.

I did so.

'What did you say it was opposite to?'

'The new Armed Police Lines!'

I gave up. 'For God's sake send someone quickly!' I yelled, 'or else there won't be anything left to save!'

After this frustrating conversation, I rushed out, grabbing a sheet from my bed on the way, plunging it in water and racing to the gate. The fire was still an alarming sight, but not quite as bad as when I first saw it. The women and children were standing around wailing, while the boys with a number of men from the Armed Police Lines opposite, were busy putting out the blaze. I counted heads. They were all there. Even the puppies and kittens, that had been sleeping inside the quarters with the boys and their families, had been saved. Some of the staff had managed to salvage bits of their clothing, but most of them had lost everything.

Strangely enough, Manikam who got out first, reported loftily that he and all his belongings were completely safe, unscathed, unhurt, untouched by the demon fire! I must admit to finding that odd upon later reflection.

After twenty minutes, by which time all that was left was a little smouldering wood, the fire brigade arrived. They had lost their way and gone to the wrong Police Lines because their telephone operator had been unable to understand the lady with the peculiar accent!

Little by little the whole story unravelled. Kundan's mother had been the first to detect that something was burning. She alerted the rest of her family. When Kundan tried to open his door he discovered that it had been bolted from the outside. So had all the other doors of the other huts! Fortunately, the wood was flimsy and they had all managed to break down the doors. When they rushed to the well to fetch water, they discovered that the bucket which was normally kept there had been taken away and hidden. Had it not been for our neighbours in the Armed Police Lines, things might have been much worse.

The fire did not make Manikam look good, I'm afraid. I never did articulate my suspicions to him or any one else, but I had this unhappy feeling that Manikam had lost his balance over his jealousy and had perhaps become one of our more dangerous inmates. I was also not willing to rule out the possibility that the fire had been caused by the three night-visitors from some days ago, though I found it difficult to attribute a motive to them as nothing had been stolen.

Like the theft of my money, which had been much talked about for a few days, the 'night of the fire' episode was soon forgotten and remained an unsolved mystery.

And then one night it happened again. Another fire in the huts. I had a dream-like feeling that all of this had happened before, which, of course, it had. We all went through the motions quite mechanically, with much less panic this time.

After this second fire, Prem decided that he had had enough of the pressure of competition with Manikam. It was one of the saddest days of my life when Prem came up to me and announced with folded hands: 'Miss-sahib, please forgive me for any mistakes I might have made in the years that I have served you. It is now time for me to leave . . .' I tried to reason with him and implored him to stay, but he was adamant. I looked into his anguished, wet eyes and knew I could hold him back no longer. Though I tried, I regret that I could not resolve the issues between Manikam and Prem; with no substantive proof of wrongdoing, I could certainly not sack one to keep the other.

So, Manikam's main rival had left us and I hoped that things would fall into a smoother pattern now. And so they did, for a while at least. Vigin took over my cooking and life was reasonably peaceful.

One day, on returning from an ambulance round, I was informed to my utter amazement that Manikam too had left! The story was—Kundan had borrowed

Manikam's cycle pump to fill up the tyres of our cycle-rickshaw, but had made the fatal mistake of not informing or asking Manikam.

On hearing this from a third party, Manikam had charged at Kundan and had snatched the pump from his hand.

'Do you know who this belongs to?' he had shouted.

'Yes.' Kundan had replied. 'I know that it is yours. I borrowed it for blowing up the tyres for the Miss-sahib!'

'And this is what I do if anyone borrows my pump,' Manikam had cried dramatically, breaking the pump across his knee.

Everyone had stared at him in amazement. He said no more and rushed out the gate. Late that night, apparently, he crept back in to the shelter and collected his belongings. But no one saw him come or go. Just, I suspect, as no one had seen him when he set fire to the staff quarters. Or when he set fire to Prem's sense of self-worth. Or when he set fire to my belief in my ability to distinguish a scorpion from a human.

Ten

THE 'PAGAL MEMSAHIB'

The small Fiat car which I had brought back with me from England in 1963 began to show signs of over-burden. Parts were perpetually breaking or getting worn out and it wasn't always possible to find replacements for that particular model. The gears were perpetually going out of order, and driving the car became as hard as pushing an unwilling elephant. It was soon clear to all of us at the shelter that our Fiat would have to be put on a retirement plan. Plenty of sunshine and ample rest.

We began to despair of ever having proper transport again. We had to cart off our sick dogs and other animals to the SPCA hospital daily for medical treatment either by our cycle-rickshaw or by taxi. We had established an understanding with the local taxi stand and it was no uncommon sight to see Miss Crystal Rogers accompanied by an assistant, escorting groups of dogs, cats (and sometimes even a goat!) in the back of the taxi.

The SPCA hospital was at least five miles away from our shelter. This exercise was not only expensive but also quite tedious and although Dr Sharma of the SPCA gave us his very willing help, I longed for the day when we could manage to have a veterinarian of our own.

We were now getting fairly well known amongst the concerned population of Delhi. Our telephone was constantly summoning us to accident cases which we never refused, no matter how far away they were, or what difficulties were entailed in getting there. The staff was magnificent and all rallied round no matter what time of night or day it was, no matter whether it meant travelling by cycle-rickshaw, taxi or even bus.

One friend who was an enormous help to us during this period was Brigadier Joyce Stags of the Army Nursing Service. She had recently bought a new Fiat—a size larger than my own—and lent it to us on several occasions. I am sure she must have had a rude shock upon seeing a photograph in our newsletter showing a donkey staring placidly out of the rear window of her car!

Finally, in 1967, we were blessed with the gift of a Morris van from one of our overseas sympathizers, a Dutch lady, Miss C. Lindleyer, who having embraced the Hindu religion preferred to be known by her Hindu name, Sister Chandramani.

I went down to Bombay myself to collect the van when it arrived, this time, I thought, armed with all the correct papers. Once again, however, I was too optimistic! The permit, despite many reminders, had only been prepared just moments before I actually left Delhi. The van, however, had been shipped some weeks earlier. When I showed my permit it was found to be not in order since the magic words 'already shipped' did not appear anywhere! Once again, letters and telegrams began to travel between Delhi and Bombay while I stood by helplessly and watched the new treasure sailing away to Pakistan.

But as I have said before, by now I had learnt the art of patience. I remained in Bombay and waited for the ship bearing my cargo to return to Bombay, which it eventually did. I finally drove off in the new car with a hitchhiking Sudanese student I'd picked up on the way, a baby crow and a sick kitten for passengers.

The journey was very exacting. We encountered heavy rains along the way and in one place had to leave the road and drive across a rushing torrent which did no good to the new tyres. It was also difficult to find rooms at the dak bungalows along the way in Nasik, Indore, Gwalior: owing to a religious festival they seemed to be all full. The result was we had to sleep as best we could under very rough conditions. My passenger was obviously not enjoying the journey and I suspect he found me quite tiresome—I don't blame him—as I stopped at frequent intervals to pick up sick creatures by the wayside. The last straw, as far as he was concerned, was when I 'wasted so much time' taking a kitten to a vet on the way.

From that moment onwards, my human passenger ceased to speak to me. Altogether, it was only Kaka, the crow, who kept up the morale of the party with his constant cawing. The car was quite full by now with a mixed crew of feathers and paws; the windows could not be rolled down for obvious reasons and the car had started to acquire the smell of a zoo. However, Kaka, unmindful of the tension between the driver and the passenger in the car, was determined to be vocal and happy. It was a shock which I shall not quickly forget, when at a halt at Agra, one of the puppies managed to get hold of Kaka and killed him instantly.

The welcome I got at the shelter upon my return had to be seen to be imagined. I gave my call signal on the horn and everyone came running—staff, their relatives, children and of course dogs and cats! Even our three sheep joined in the general commotion of welcoming me. The car was admired and a short ceremony, consisting of an *aarti* and the garlanding of the bonnet, was performed to ensure its longevity and good health.

After trying out a series of young and inexperienced drivers, none of whom lasted long, we finally obtained the services of an older and more cautious driver, Pritam Singh. He turned out to be a very dependable addition to our staff, always ready to take any extra driving load, never complaining about late nights or early mornings. It seemed as if, for a time at least, my life might be a little easier.

Physically, yes, it was. Mentally, it wasn't. Now that I had a little more time to reflect, I revisited the huge issues I was up against. There was always an underlying heartbreak when one considered how much there was to do as far as animal welfare was concerned, and how little actually was in one's power to remedy.

I had been asked by Rukmini Devi Arundale, our first president and still our chief patron, to compile a humanitarian handbook which the Animal Welfare Board was publishing. On the way to Madras to do this, I stopped for a few days in Hyderabad to take the advice of the Rev.

E.C. Early, a retired missionary who was a great champion of animal welfare. We met in the house of the local SPCA Secretary, Mrs Brij Mohan Lal, a generous lady who was also greatly committed to the cause of animals.

From Hyderabad I went on to Madras where I sat for many days composing the text for the handbook and weltering in the accounts of cruel practices which poured in from every corner of India. Animal sacrifices, the transportation of animals, slaughter houses, the different ways of killing a pig, each more hideous and diabolical than the last.

It was not until I undertook this particular task that I realized the full horror and exent of the atrocities being practised in the country in which I was living. When one casually reads about cruel treatment to animals—it stings one, no doubt, but the pain is only momentary. The inhuman practice against animals in Borneo or the Philippines; the slaughterhouses in the West; the specialized farms around the world where animals are born, fattened and nurtured only for the purpose of being ultimately dished out as a main course on an elegant dining table—these acounts leave one with a faint, nagging guilt for being a silent participant in this horrifying and violent drama. But it is only when one dedicates oneself to understanding the entire depth of this gory business that one comes face to face with the truth: the human race is appalling in unimaginable ways. Is this what evolution was all about? Is this what is known as the 'refined human spirit'? Lord help us if it is.

I recall an incident that took place on a chilly February evening. The phone at the shelter rang and I answered it.

'The Animals' Friend,' I said.

The voice that came over the wires was of an old friend of ours who was a known and respected animal-lover.

'I'm terribly distressed,' he said. 'I've just seen a dreadful sight and I don't know what to do.'

He went on to say that on his way home a few minutes previously he had heard a great commotion and seen a great crowd of people. On coming closer he had found that the object of interest was a pig, which, with its mouth and feet tightly bound, was having its bristles slowly pulled out by a hefty young man. The pig was screaming in agony and terror, while the watching crowd laughed in great amusement. Finding himself heavily outnumbered, our friend felt the best thing he could do was to telephone for help.

Unfortunately, it was a Sunday and our van was already out on its evening round. I borrowed a jeep from a visitor at the shelter and dashed off to catch hold of an SPCA inspector, which, again since it was Sunday, I was unable to do. However, going to the nearest police station I enlisted their aid and drove on towards India Gate where we were to meet up with our friend, the informant. The police, upon hearing that there was a large crowd, gathered together a few more constables from a nearby area and off we went. Unfortunately, by the time we reached the location the crowd was gone. So was the pig, and the bystanders wore looks of innocent surprise. It was clear that the people who had organized this show had had a presentiment that trouble was at hand and had sensibly decided to wrap up the show for now.

But it was not long before the police unravelled their names and addresses and brought them to book.

Pigs are perhaps the most to be pitied of all animals in India. There are, in all, five ways of killing them, each one more painful than the last. Most of the killing is done 'unofficially' by those who breed them. This is generally done under the cover of night and there seem to be no restrictions upon the manner in which the animals are killed.

Pig bristles fetch a good price. They are considered superior if they are taken from a live animal. A boar bristle brush is a fashion statement. Where will our vanity take us? How much suffering can we cause in the name of fashion, superstition, tradition and culinary desire?

Anyway, I continued sorting and collating material for the handbook in a mixed state of despair and rage. How was one to stop these perversions, these cruelties . . . ever? It was like pitting one's puny strength against a solid mountain or a raging torrent. Would the public at large, in this land of *ahimsa*, never revolt against these horrors and do something to put an end to it?

In Madras I enjoyed the hospitality of Capt. and Mrs Sunderam, the founders and organizers of the Blue Cross of India. The organization was a family affair in which every member took his or her part and performed their duties with devotion. I greatly admired their efforts.

It was a joy to see the active dedication and compassion of the Sunderams, a stark contrast to the apathy that I had encountered in so many people. What could one do to shake people out of their apathy and get them to act? The harder one tried to arouse public sympathy, the more one realized that one had barely pricked the surface.

*

I have always been deeply moved by the sight of a lost dog. I do not refer to the stray dog which was born in a ditch and probably never knew the love and caring of a home, but the dog which has been deliberately abandoned because the owner does not want it any more.

I shall never forget the case of Constance. It was a cool pre-winter afternoon. I was chugging along on one of Delhi's main roads in my autorickshaw, enjoying the magnificence of the Purana Qila, when I noticed just ahead of me a well-dressed man in an autorickshaw and following behind him an already exhausted looking small black bitch. She was straining every nerve to catch up with the rickshaw. I also noticed the man leaned out of the rickshaw every now and then with a mixed expression of anxiety and relief as the vehicle gained distance on the tired little dog.

Suddenly the three-wheeler turned a corner and after it went the panting dog while I followed in my autorickshaw. Having witnessed on many occasions the behaviour pattern of humans and animals in a situation where the master is abandoning his pet, it seemed to me quite clear what the roles were here. The man in the rickshaw was quite surely the master who, for his own deranged reasons, was washing his hands off the little bitch, who had been quite surely his faithful pet till this cruel day.

The chase continued for quite some time: I marvelled at the little creature's stamina and I cried from within at her hopeless situation. To realize that you are unwanted by the people you love and trust with all your heart must be the most devastating experience in the world.

The scooter turned a second corner, the dog went racing after it. As she was trying to double her efforts to catch up with her master, which was what I by now assumed him to be, two large dogs rushed out of a nearby gate and blocked her progress. Her look of horror, of despair, of disbelief at what had just been done to her, was heart-breaking. It was at this point that I succeeded in catching her and taking her back to the shelter. There, she refused to eat for several days but sat all the time watching the gate, while the sound of a distant scooter was the only thing that set her tail wagging. She was obviously waiting trustfully for her adored master to come and reclaim her. She had very little time for any of us. All she wanted was her master—the master who never came.

We loved her nonetheless and named her Constance. An appropriate name. She showed us how pure and unwavering an emotion loyalty can be.

Another instance which comes to mind is the case of the honorary *ayah*. For two or three years she was the nanny to a family who lived in a posh colony of New Delhi. Every day she was the one who escorted the children to school and back, guiding them safely through the traffic and seeing that the smallest one did not stray off the pavement or fall down. She loved the whole family and managed her role responsibly and with devotion. One day the family moved away—but made no provision for their honorary *ayah* who was left behind: after all she was only a dog, a rather fat golden labrador. She would doubtless manage somehow.

But Soni the labrador's heart was broken. She had loved them so much and now she couldn't understand why they didn't love her in return. Beaten, hungry and homeless, Soni found her way to the Animals' Friend shelter. We advertised her in the hope that she would find a caring home. It took several months before that happened and in the meanwhile Soni discovered that she could continue her duties at the shelter as chaperone and companion to the children who lived in our staff quarters. She served them with utter dedication and they responded by giving her immense love. It is hard to come by a faithful heart and more often than not we humans tend to ignore the presence of a pure soul amongst us.

There were many people who thought I should not be wasting my time and energy upon animals but giving these to the cause of humanity. 'We think humans are much more important!' one welfare worker remarked, turning away in disgust when I rattled my near-empty collection tin before her during Animal Welfare Week.

Certainly, humans are important—to me all creatures that breathe and walk this planet are important, but this has never been understood. My work for animals has brought me an undue share of criticism—which is less hurtful when it comes from strangers, but goes deep when coming from friends and relatives.

A relative with whom I was staying some years after I started my work in India handed me a letter across the breakfast table coming from a mutual friend. In it she described her daughter's successful career. 'What a pity,' her letter went on to say, 'that poor dear Crystal never seems to have found her niche.'

I handed the letter back with a smile.

'Well, Mr Wilson certainly seems to think that I am truly wasting my life!' I said.

My relation, the kindest of souls, leaned across the table and said with the utmost sincerity, 'My dear girl, I couldn't agree more.'

On more than one occasion the same relative would stop someone who was in the middle of recounting details of some bus or train disaster, in which a number of people, including children, had been killed.

'Oh, don't tell Crystal!' he would interrupt, 'she won't be in the least bit interested unless the bus was full of Pekinese!' His jibes were probably meant as jokes and I am sure he and others who made similar remarks did not understand that sometimes even jokes hurt.

I could quite understand being dismissed as a '*pagal* memsahib'—mad lady. That was inevitable, especially if someone had seen me on an occasion when, on a busy Delhi road, I was wriggling under a stationary bus on my tummy, in order to rescue a terrified little mouse which had been thrown there by a sadistic little boy.

I have had numerous insults dumped on me: so much criticism and so many jibes for what I considered to be my sacred mission, namely the rescue of abandoned animals. I agree that abandoned animals do not fare high on the protocol list of 'animal activism': a far more visible and 'important' exercise would have been to organize a society that would have been vocal in its opposition to cruelty against the beasts of God. Whether that society actually managed to do any good would have been beside the point. People, it would seem, are more interested in words rather than action. I shall quote some of the abuse that has come my way: 'What a pity to spend so much time on animals when human beings are so much more important!'

'The Animals' Friend was founded by a mad Englishwoman who wastes her time running about all day long chasing street dogs!'

'Don't send any animal to the Animals' Friend Shelter. They kill them there …'

A snatch of conversation:

She: I've read letters in the newspapers that indicate that the shelter is doing good work.

He: Ha! Well, you should know that most of those letters are written by Crystal Rogers herself.

No, being called a *pagal* memsahib did not worry me in the least. What did cause me great heartache were the numerous times that I had been let down by my allies, people I thought were like-minded, who I believed had the same passion and commitment to the work as I did myself. This I found extremely disillusioning and it saddened me terribly when, one day, a friend and colleague remarked, 'Why are we chasing after people now? I thought our job was with the animals. There are enough organizations to take care of suffering humanity: the American Red Cross, Mother Teresa, UNICEF, and literally dozens of societies. Must we take on this mission as well?'

This was said after we had taken a group of Tibetan babies from the Buddha Vihara refugee settlement; all were suffering from pneumonia and would easily have died without immediate medical attention. I was shocked by my colleague's statement.

One thing is certain: one can never please everybody and those who allow themselves to be upset by criticism will never stay the course. I have, over the years, built a wall around myself and have become indifferent to the barbs, to the criticism, the sneers and the jokes.

To me compassion is all-embracing. It can not be stuffed into a single compartment. I have never been able to see how it is possible to feel compassion for an injured kitten or dog and not feel the same for a sick baby or a destitute old woman. Or vice-versa, for that matter. If the *pagal* memsahib can make a small difference to any living creature's life with her so-called madness then that is the way I choose to be.

Eleven

ON THE BRIGHT SIDE

I don't want anyone to run away with the idea that my life at all times resembled the trials of Job. There is always a brighter side to everything and a sense of humour does a lot to get one through some of the bad patches. We had our funny moments and were thankful for them, whether the causes were animal or human.

For chasing away the blues I cannot recommend any animal more highly than the monkey. Our series of monkeys had always been the clowns of the shelter. Whether riding on the backs of other animals, admiring themselves in front of a hand mirror or draping themselves in bits of rag and strutting around as if on a modelling ramp, I defy anyone to watch a band of monkeys for long without brightening up.

Some of our human pick-ups had also, over the years, provided us with interest and amusement—and sometimes pathos as in the case of the naked lady. I found her sitting at the side of the road one morning, devoid of any clothing, breaking one stone with another and humming all the while. The nights were still cold and I wondered how she was managing to survive in that condition. I attempted a conversation with her but she just looked through me as if I was thin air. I went back to the shelter and got Kundan, thinking he might be able to overcome language barriers, if any, though I might add that my Hindustani is quite passable. With a face suffused with blushes and eyes screwed into the ground, poor Kundan tried to win her confidence.

Within a couple of minutes Kundan had established that she wanted to go to Jaipur.

'Oh! You really are very lucky then!' he exclaimed in a voice full of feigned delight. 'Jaipur happens to be just where this car is going. If you will get in, you can be driven there.'

The naked lady appeared to think this was a good idea. She picked up an indescribably filthy rag out of the nearby gutter, wound it round her head, then, rising with great dignity, stepped into the car. Once seated beside me she started to sing with tremendous enthusiasm. The song was on three notes but contained only two words: *Hat* and *Babu* which when translated means 'get out of my way, sir'.

Hat Babu! Hat Babu! sang the naked lady lustily, as I drove her through the streets, an object of curiosity to many interested eyes and ears.

To my relief the Poor House was not far off.

'We will just stop at this hotel for a meal,' said Kundan with such naturalness that I could have almost believed him myself!

Our passenger got out without demurring and sat in the veranda while I managed to persuade the superintendent to take her in. I somehow felt that she might be a little bit unsuitable for our shelter. I could not possibly subject poor Kundan to days and weeks of bolting his eyes to the earth and going around with permanently rouged cheeks!

Perhaps our most amusing pick-up of all was the 'Prime Minister's cousin' . . .

A young man living near us, not over-blessed with brains but good at heart, called us one morning requesting for the ambulance to be sent to his home in Karol Bagh, about five miles from us.

'What do you need the ambulance for?' I asked.

'There is a lady here in great distress. She is practically without clothes, has no money and nowhere to go.'

Imagining a situation similar to that of the naked lady, I was duly sympathetic. The ambulance was dispatched and an hour later she arrived. It was a cold winter morning and here was this elderly lady in the thinnest of georgette saris—quite unsuitably dressed for the biting cold winter of Delhi.

'Good morning. I have only just arrived from Bombay and I have no money.' She spoke in fluent English. Her forthrightness was quite stunning. I decided to be equally forthright.

'Didn't you bring any warmer clothing with you?' I asked, noting that her only luggage was a tiny tin box about six inches by eight inches.

'No,' replied the shivering woman. 'You see, it was quite warm in Bombay.' She continued conversationally, 'Don't you have another cardigan like the one you are wearing?'

With one of the many humans who Crystal helped as part of her compassion programme

At the new shelter land in Jaipur

Crystal with two donkeys and a dog in the Jaipur shelter in the early 1980s

Help in Suffering, Jaipur: with a human patient

Receiving the RSPCA Richard Martin award in 1978

The first CUPA shelter (1991)

Bangalore, 1990: Crystal aged eighty-four

Jennifer Butt, International Director of the RSPCA, with Crystal in 1996 a few months before her death

Bemused, I walked into my room and pulled out a cardigan from my cupboard. She put it on, but since she was about half my size, it didn't fit her.

'Haven't you got anything smaller?' she demanded.

I pointed out to her that she was about a foot shorter than me and would have to make do with what she had been given. The next thing on the agenda, I presumed, would be to feed her. I asked her what she might like to eat. She said she was on a fast.

'Even on a fast, I suppose you do take *something*!'

'Well, since you insist . . . I would like a glass of milk and perhaps an egg or two.'

'How would you like the eggs done?'

'One to be poached in milk and the other one fried.' She adjusted herself in the chair and then went on, 'Have you got any powder milk? I should like some and also a banana with a little black pepper. Please.'

Finally all her demands were met and except for the fried egg she dumped everything into the milk and stirred vigorously. I couldn't help wondering what she ate when she was not fasting.

I asked her why she had come to Delhi without any money.

'I thought I should be staying with the Prime Minister. She is a relative of mine, you see. I knew I should need for nothing. She is such a sweet and kind person. But when I arrived at the Prime Minister's residence they told me that she was away in Europe.'

'Where do you propose going, then?' I asked.

'Where can I go? I have no money!' She looked at me

meaningfully.

I tried to put her off by telling her that I had no extra bedding, no spare rooms, nothing. She stopped me in my tracks by insisting that she had no need for anything of this sort since she was currently spending the nights in meditation.

'But it would be very uncomfortable for you,' I persisted. 'The dogs bark at night and you would find it very disturbing. Besides, as I said, I don't have any extra room.'

'That wouldn't matter,' the PM's cousin dismissed all my objections with a wave of her hand. 'I could sit in the kitchen on a chair.'

And so this situation continued for the next few days. The staff was getting absolutely frustrated with her eccentric diet, her constant demands for blankets, pillows, sheets, and her haughty attitude—on account, no doubt, of being 'related' to Mrs Gandhi! And, I must admit, that it wasn't quite easy for me, either. She was, to put it very mildly, exceptionally irritating. One morning I decided that the poor sap, known to all as Crystal Rogers, must take a firm stand.

I went up to her and announced that I would be putting her in the care of social workers. She interrupted me, 'But you see, I'm not feeling very well today.'

I resisted the temptation to tell her that anyone upon her diet would end up in a hospital, in any case.

'I know one of these social workers personally and I am sure they would be delighted to put you up until Mrs Gandhi returns.' I was cold as stone. 'Now, if you will please be ready in ten minutes we shall go on our way.'

She got into the car with considerable reluctance. As I was stepping into the car, Vigin whispered into my ear, 'Miss-sahib, please don't give her any money. I have seen two thick bundles of notes stuffed inside her blouse.' This, I must say, got my mercury levels soaring. There is one thing that I feel should be punished severely and that is lying. I was even happier with my decision to oust this woman from the shelter now with this new piece of information.

During the journey the 'cousin' maintained a dignified silence. When we reached our destination, I jumped out of the car, came onto her side, opened the door for her, pulled her out as politely as I could, folded my hands in a *namaste* and spoke fast, trying to leave no scope for a debate at this stage.

'Goodbye. I know you will be well looked after here, till the time comes for you to go to the PM's residence.' The lady was too outraged to reply. I bowed and quickly backed into the car and sped off like a maniac, not looking back even once for fear that I might see the 'cousin' racing after me.

*

Over a period of time I found myself gaining the reputation of an agony aunt cum mentor cum expert on foreign affairs. A number of youngsters from middle-class Indian families would approach me for guidance on how to conduct themselves abroad, if they ever got the opportunity to visit a foreign land, which seemed to be an all-absorbing dream for that generation. I decided to put together an informal presentation on general etiquette for this purpose. The chapters dealt with a few basic aspects of social behaviour: manners on a journey; manners in a hotel or home; manners in the dining room and manners at a party.

Out of this last chapter I will quote:

> Perhaps one of the most important things at any social function in the West is to show yourself to be a good and attentive listener.
>
> I hope I may be forgiven for saying that up to this point I have not found this quality in most of the Indians whom I have had dealings with. But then, possibly I have not said anything that they have found sufficiently interesting!
>
> Sitting at the breakfast table with a charming Indian family the conversation is likely to go something like this:
>
> My hostess: How did you enjoy the party last night? Did you meet anyone interesting?
>
> Self: I enjoyed it very much. There was a most unusual man there who . . .
>
> Hostess: Don't take another apple, Bablu. You have become so greedy. You haven't even left one for Auntie . . .
>
> Bablu: But apples are my favourite, Mummy!
>
> Self (*hastily*): It's quite all right. I'd rather have a banana. As I was saying—this man I met yesterday came from Lhasa. He told me . . .

My host (*to the hostess*): Are you wanting the car this morning?
Hostess: Not in the morning. I'll need it for shopping in the afternoon.
Host: I'll get the driver to bring it back after lunch.
Hostess: Okay, *theek hai.* (*Turning to me*) Do go on. You were saying?
Self (*talking very fast*): Yes, so this man told me the most extraordinary thing. He said he could see all of my past, present and future and that ...
Bearer: Memsahib-*ji*, the darzi has come. Shall I tell him to wait?
Hostess: Yes, of course. I had asked him to come a week ago and the silly man turns up today. I need him to make a coat for me.
Host: Another coat? Didn't you just have one stitched?
Hostess: I don't know what you are talking about. I haven't had any made all of this winter.
Host: What about that red one?
Hostess: That, darling-*ji*, was not a coat. It was a jacket!
Self: (*keeping quiet but following this backward-forward dialogue like one follows the ball in a ping-pong match*).
Hostess: Sorry, Crystal. Please continue, I'm listening.
Self (*a bit defeated by now*): Well, he had come down from Lhasa and claims to have these extraordinary powers to look into the past and the future; he had the most remarkable stories about . . .
Host's mother: Why doesn't somebody put on the fan? It's so hot in here!
Host: I don't call it hot. You must have a fever if you think it is hot today.
Hostess: Yesterday was exceptionally cold. Today it is

cold, but definitely not warm. But if you want the fan . . . bearer! *Punkha chalao*!

(*Everybody looks up at the ceiling for a few moments*).

Hostess: So, about this man. It sounds so interesting. Please tell me more.

Self (*giving up the unequal contest*): I'm afraid I can't seem to remember. It's hardly important anyway.

*

The above may sound like a tall story, but I can only say that conversations like this have taken place in my experience again and again! So, if you want to make yourself popular in the West, listen without interrupting.

Also be careful about the nature of questions that you may ask. Questions upon generalities are quite legitimate, such as, 'Have you lived in this part of the country long?'; 'Have you ever visited India?'; 'Have you seen the film at the Plaza?' Questions to be avoided are personal ones, such as, 'Are you married?'; 'How old are you?'; 'What is your income?'

Such questions would be considered not only inquisitive but rude. Finally, be sure to know when it is time to leave. It is better to look at the clock and remark, 'Well, I'm afraid I must be going. It's quite late.' It's infinitely better to be pressed to stay on rather than to sit on and on, long after the conversation has started flagging and as your hosts are trying their utmost to suppress their yawns.

Twelve

GAINING POUNDS

Back to the shelter. Our constant problem remained that of funding. By the end of 1967, it was clear that I should have to go away on another, longer, fund-raising tour, or The Animals' Friend would cease to exist.

I finally left by sea in March 1968 and as I had to go all round the Cape of Good Hope, on account of the closing of the Suez Canal, I did not arrive in England until the end of April. I had no set program but had brought a number of fresh films with me, which this time I took great care to see did not get lost or stolen.

Once again I made a beeline for Ida's house which, since the death of my mother, was a home away from home for me. As usual my two dear friends, Pauline Skeate and Andre Bannerman Phillips, who had taken on the jobs of our UK representatives for the central and coastal regions respectively, rose to the occasion magnificently. While Pauline dealt with the enormous amount of mail which kept on pouring in, Andrea arranged meeting after meeting for me along the south coast. Staying with her in her caravan, the night before a meeting I would hear her moving around in the early hours of the morning in her tiny kitchen, making sandwiches, icing cakes and wrapping up little bags of toffees and sweets. Never was there a more tireless worker for the cause.

At the beginning of the trip, I had, on Andrea's advice, invested in a second hand Morris Minor which proved very useful when it came to touring around the country. The car's number plate bore the letters ELJ and I wasted no time in christening it Elija. With Elija's help I toured England successfully, from Carlisle in the north, where Alfred Brisco, organizing secretary of the National Equine and Smaller Animals Defence League, had arranged a meeting for me, to St Ives in the south, where Miss Raymont, a new member of The Animals' Friend, had done the same.

Back in London, I decided to revisit the College of Psychic Sciences, having been a member for more years than I care to remember. The college has a fine library dealing with matters of the metaphysical and supernatural kind. Since the loss of my dear and precious friend, Jim, in World War II, I had been drawn towards the afterlife. I would not call myself a 'spiritualist', which I am inclined to think is a misleading title, nor have I ever been in the habit of running to mediums to ask if my deceased Aunt Fanny or Uncle James would approve of my next move. I am not even sure if Aunt Fanny or Uncle James would be particularly interested to answer questions about pregnant donkeys and hyperactive hens! But once having glimpsed the promise of the life that beckons from 'the other side' it is difficult to turn one's back and walk away as though nothing has ever happened.

After some days of drifting in and out of the college library I was invited to a sitting by a Mr Bo Goram, a Swedish medium, at that moment quite unknown to me. He instantly picked up on my background: he mentioned the fact that I had just come from a 'hot country' and wondered why he saw me surrounded with animals.

'There seem to be a large number of dogs, but not very nice looking dogs. Most of them look thin and ill.'

I smiled and accepted the statement, but said no more.

After an amazingly correct reading, during which he sensed that I was 'collecting money for a something', he suddenly came out with the remark:

'Are you going to America or Canada?'

I replied that I was not.

'I think that you will,' Mr Goran said firmly. 'In fact, I am quite certain of it. I think you will go to both countries, but it may not be immediately . . . within the next two to three years.' I made some noncommittal noises since I could not see how I would ever have the money to go to America and Canada.

At the end of the sitting I thought it only fair to let him know that he had been accurate on many counts. I told him that I was running a shelter for animals in India, which explained the hot country and the animals that he had envisioned.

Bo Goran expressed much interest. He asked me whether I knew a Miss Ruth Plant.

'No, unfortunately not.'

'Well, you should,' he said. 'She too is interested in animals and I am sure she could help you. Do write to her.'

I agreed to do so and took down her address. On an impulse I did send her a letter, little knowing how far-reaching Bo Goran's advice would be.

A few days later I got a reply from Miss Plant asking me to meet her at a restaurant in Knightsbridge.

Waiting for the lady at the appointed hour in the restaurant, I sipped my tea and made small talk with the gentleman at the next table. Some moments later, Ruth Plant walked in and seconds later we were joined by a friend of hers. In all the confusion of the moment, the shaking of hands, the pulling of chairs, the ordering of tea, I could not catch her friend's name. Anyway, the conversation began and we soon discovered that we were all quite comfortable with each other. The friend informed me that she was arranging an Animal Welfare Conference at Attingham Park in June. I had never heard of Attingham Park and had no idea where it was but was ready to go anywhere.

'I'd like you to speak at the conference and show your films, if you can,' said the lady.

I gladly accepted and on that note we three parted. At my next visit to the College of Psychic Science I wandered into a demonstration on clairvoyance being given by a medium who I had never set eyes on before. The room was quite full and I occupied an inconspicuous spot at the back.

Suddenly I saw the medium turn in my general direction; with eyes firmly focused on me she called out across the room, 'You are from abroad, aren't you?' I nodded quickly, having been caught off guard. 'Yes, I thought so,' she continued, 'are you collecting for some cause?' Once again, my natural scepticism in these situations rose to the surface and I convinced myself that this medium had possibly met Bo Goran and had got some general information out of him. I smiled back at her politely. 'I see you going to America and Canada, both countries . . . The amount you collect here will be only a flea bite compared to what you will get over there!'

I thanked her dubiously. Just as I thought she had finished with me, she turned back.

'Do you know anyone by the name of Hodgson?' she asked.

'No. Not at present.' I replied, thinking of a Colonel Hodgson I had known several years ago.

'Well, remember the name.' The medium said briskly, 'I think you will find someone of that name will be of great help to you. Hodgson—living in Shropshire.'

It was not until I met Ruth Plant again, a few days later, that I enquired the name of the friend to whom I had been introduced at the restaurant in Knightsbridge. She was a Miss Hodgson and did, indeed, live in Shropshire!

The conference at Attingham Park, which I attended in June, turned out to be an enormous help to me. I not only made new friends and recruited more members but also came away with over one hundred and thirty pounds in donations.

One of the people I met at the conference was Elma Williams, the writer, whose delightful book, *A Pig in Paradise*, immediately caught my attention as it lay on a table outside the conference hall. Since she and I were two of a kind, it was inevitable that we should click. She invited me to visit her in her Valley of Animals, which I managed to do some time later.

This lovely place, which Elma called 'home', was in a valley in North Wales overlooking the sea. As I stood before the entrance, breathing in the sharp, salt air and watching the white gulls spiral over the grey-green waters of the Irish Sea, I felt like this was a sort of homecoming for me.

My stay at the Valley of Animals with Elma, her friends and her variety of animals was completely joyous and I should have liked to stay there longer except my own home, back in the plains of India, was beckoning and it was time to return to reality and hard work. But I was returning with a song in my heart. Over the past few months I had managed to raise over two thousand pounds for the Animals' Friend shelter, not a mean sum, not something to sniff at, no indeed, two thousand pounds and more. Well done, you old cow, I told myself as I packed my bags for the journey home.

Thirteen

FRUITCAKES AND TEA

It was a relief to get back to India and to the shelter. From the moment of my arrival my indigestion and bad nights—that had been troubling me throughout my stay in England—ended magically.

At the very first committee meeting of The Animals' Friend I proposed it was time for us to have our very own veterinarian on the spot, to deal with the many emergencies that arose. Dr Sharma, veterinary officer-in-charge of the SPCA, who had first been with us as our advisor and later as our Vice President as well, had done his best to deal with the shelter's problems. However, he lived a distance of five miles from the shelter and in the heat of the summer, the continual runs between the shelter and Dr Sharma's were getting to be far too stressful for both us and him.

The committee accepted the proposal and Dr M.L. Bhatia, a retired veterinarian, was engaged to be at the shelter daily from 9 a.m. to 5 p.m. This came as a great relief to me: finally our patients could get medical attention as soon as it was required. We stocked up on all possible manners of bandages, medicines, vaccinations and soon had a full-fledged dispensary up and running for our four-footed friends.

Meanwhile, a new development had occurred. We had been asked by the Delhi Development Authority to shift out of our present orchard to the adjoining one. We agreed since there seemed nothing else to do: our present land was being taken over for the construction of a road and quite obviously our complicated domain could not possibly sit in the middle of a busy thoroughfare. So, again, we had to shift the whole caboodle, but this time, mercifully, only a few yards away. The new land was officially ours, but before we had taken it over the fruit from the orchard had been sold to a third party. These people made life impossible for us. They were interested in marketing the fruit and our attempts to take our large animals in to graze were continually impeded by them; in fact, they denied hotly that we had any rights to be on the land at all. And this after being shown papers and proof that we, the shelter, were indeed the rightful owners of the land!

Our young goatherd came to me one morning in utter perplexity:

'What shall I do, Miss-sahib? You ask me to take the goats into the clearing next door but when I get there the fruit-pickers threaten me and say the goats will spoil the fruit. They told me never to step foot on that part of the land, otherwise they will do something horrible to me.'

'Don't worry,' I told the youngster, controlling my temper as best as possible, 'the land is ours. They can do nothing to you, I promise.'

Even as I said these words to him, I knew that I had made a false promise.

My own goat Melani, who was pregnant at the time, was taken seriously ill the same night. Kundan and I tried our very best to diagnose her problem and help her through the night. By early morning I could bear her moans no longer and putting her in our little van-ambulance, we drove her to Dr Bhatia's house as fast as possible. Although asleep when we arrived, he made no delay and quickly attended to her.

For several hours we battled for Melani's life, till finally just a bit before 7 a.m., Melani breathed her last after producing a premature kid who died within minutes of his birth. Dr Bhatia noticed a large bruise on the kid's head and determined that Melani's condition must have been the result of a deliberate and violent attack. It did not take much imagination to understand who might have done this.

My life seemed to escape out of me as I saw my dear Melani lying there cold, eyes gazing into the eternity that is death, and her premature little kid next to her, his life stolen from him mercilessly by one cruel act. From that moment onward it was open warfare between our boys and the fruit-pickers.

The abuses flew openly across the orchard, between the enemy camps. Time and again the boys from our shelter and the fruit-pickers had to be pulled apart by neutral parties to prevent bloodshed. Out of sheer rage and a human desire for revenge the boys would sneak into the other part of the orchard and damage the fruit. I pleaded with them to put an end to this hooliganism and decided that time had come to put this issue before the right authorities for a final solution.

At last, a meeting was arranged between the DDA municipal committee who had sold the fruit, the fruit-pickers who had bought the fruit and us, who were living in daily tension because of the fruit.

We all met at our orchard. Tempers on all sides were flying high, there were hot points and hotter counter-points, accusations and defences, threats and counter-threats; finally, after witnessing this atrocious drama for several minutes I could take no more and burst out: 'Oh! For goodness' sake, shut up!'

There was a shocked silence on all sides—whatever happened to English manners? At that very moment my friend Jill arrived like an angel of peace. She led me away from the group and advised that I would gain more points with the DDA if I remained calm.

'How can I remain calm? How can I possibly be pleasant?' I growled.

Right then, Vigin announced that tea was ready.

'Would any of you gentlemen like tea?' I asked, forcing my face into what I hoped resembled a hospitable smile.

But the outraged men were not deceived.

'No, thank you,' replied their spokesman coldly. 'We still have plenty of work to do and at this rate we will not be back at the office before late evening.'

'Well, then perhaps it is even more reason for all of us

to take a tea-break?' I parried, forcing the corners of my mouth upward into a sickly grin.

And then suddenly I had an inspiration. I remembered the exquisite pink sugar cake which our President Mrs Wahi had baked for me only the day before.

'As it happens, gentlemen, today is my birthday. My hundred and fourth birthday, in fact.'

Where that particular number came from, I will never know. 'I would be very pleased if you could share the cake and this day with me.'

Reluctantly they followed me to the garden where Vigin had set an attractive evening tea on a table under the trees. Once the tea was poured and the cake sliced, the atmosphere changed miraculously! Everyone sat around smiling and congratulating me. Hearty congratulations! Many happy returns of the day! I heard myself being wished on all sides and wondered how many happy returns I could expect at the age of a hundred and four!

By this time, mouths full of pink sugar cake, everyone was in high good humour. The DDA apologized, the municipality apologized, the fruit-pickers, on hearing that they were to receive compensation, assured me that I was like their *mata-ji*, their *behan-ji*, their dear and respected friend and that I should have nothing to fear from them any longer.

Everybody was happy. We all shook hands upon promises of eternal friendship. Poor Melani had given her life for a greater cause.

Jill was right. It certainly pays to be pleasant. It also

pays to have cakes handy at times of confrontation. Above all it pays to jack up your age and celebrate a century plus of life on earth. People naturally can't disregard the wishes of very old ladies who have invited them for tea!

Fourteen

THE UNKNOWN MARTYRS

It is difficult to say which of the animals dominated by man suffer the most. The cruelties of the West are different from the cruelties of the East.

For far too many years, I have seen the barbarous methods by which animals are slaughtered in the reeking abattoirs and alleys of Indian cities. The cruelty involved in the transport of domestic animals, the terrible distress and terror of monkeys exported for medical research. The hunger, misery and bewilderment of animals turned adrift because they no longer serve any useful purpose. The agony of sick and lame draft animals who are still made to work, beaten and over-loaded.

All the above descriptions are patently obvious to those who live in the East. But do not imagine that the West is any less cruel. It is equally so. While putting up a show of being humane, behind closed doors, where none may see, the blackest and most unforgivable crimes are committed.

The West must bear the burden of guilt for the inexpressible horror of, amongst other things, killing animals for the sake of vanity. A whole fashion industry grew and flourished around the slaughter of small, defenceless animals like minks, rabbits and the red fox to ensure that ladies of wealth and social standing could wrap themselves in fur, for a cold evening out to the theatre.

Another industry to flourish in this very century has been cosmetics. The testing for many cosmetic products is performed on unsuspecting animals who are not aware that the foul, stinging liquid that is sprayed into their eyes—to test its potency—is what is used by women world over to keep their curls in place. Greater would be their surprise if they were to know that the chemical used to clean something as inanimate as an oven is what is infused into their systems to see how long it takes to corrode their internal organs.

The exploitation of animals for amusement is yet another nightmare.

'Mother, let's go to the circus, please, please!' A bright-eyed child will say.

'Oh, all right,' says the mother, showing obligatory reluctance. The trip is planned. The day comes. Mother and child pay for tickets, which if they thought a bit about it, is a ticket to hell on earth. Balloons, cotton candy, splashy banners fluttering in the breeze, carnival music. So much cheer is hard to imagine in a world so fraught with suffering.

They take their seats under the 'big top' and wait expectantly for the 'greatest show on earth' to begin. The performance will illustrate how man, the superior species, has brought the animals of the wild under his supreme control. What the spectator does not see, however, is how that training has been done.

How the whip has been cracked with lethal force on a lion's back.

How a heavy stick has been smashed on a baby elephant's rump.

How the seal plucked out of the vast, cold waters of the Pacific has been shoved into a pool the size of a child's playpen.

How the chimpanzee has been smacked mercilessly when it has shown reluctance to conform to the rules of the house.

How the spirit of each animal has been crushed and squeezed out till all that is left are the hollowed bodies of furry puppets.

The mother turns to her awestruck child and asks: 'Enjoying it, son?' He nods vigorously, mouth full of popcorn. The show comes to an end. There is thunderous applause. Mankind has enjoyed and subconsciously congratulated himself on his unquestionable dominance.

The elephants take their bows and turn to vanish into their enclosures, perhaps dreaming wistfully of a day when they can again roam the lush forests of Asia.

The lions hold their paws up to salute man and are led back into square cages with strong bars that shut out the vision of watering holes, of cool rocks, of long hot afternoons in the open savannah.

The chimps doff their caps to the cheering crowds and disappear into the darkness behind the curtain that separates the entertainment from the horror.

'Can we come again? Tomorrow?' The youngster asks,

as mother and son stroll out of the gates of hell, hand in hand, dusk gathering around them.

I have visited yet another form of hell: an experience that I should like to forget but never can. It was during the tail-end of my trip to the United States (which did materialize within the time-frame prophesied by Bo Goran and the lady medium I had encountered at the College of Psychic Studies on my trip to England) that I was invited by Dr Lambert, a recent acquaintance, to visit a vivisection laboratory . He asked if I should like to accompany him—I could think of nothing I would like less but felt it cowardly to refuse. I agreed to be picked up at 9 p.m. the following night.

I spent that evening anticipating the sights and sounds I might have to witness the next day. Not having had prior exposure to a vivisection laboratory I could rely only upon my imagination and hearsay. I recalled the stories I had read in magazines and journals dedicated to animal welfare; I had always found it difficult to believe that such horrors could actually be perpetrated in a civilized world. I tried to comfort myself in the thought that perhaps some of these stories were exaggerated and made to sound more fiendish by the animal activist lobbies to help further their own cause.

It was a wet New York night. We drove through the bustle and neon of the city until we reached open stretches of highway. By the time we arrived at the laboratory it was past 10 p.m. Dr Lambert led me in. I felt as if I was walking into some studio set for a futuristic science fiction movie. Slippery floors, low ceilings with hidden lighting, all kinds of security systems with blinking lights, blinding white walls and the kind of silence that walks hand in hand with death. A strong odour of antiseptics and animal essence wafted down the hallways. I saw a lone security guard patrolling the floor—the only sign of life.

After a few more minutes of walking we turned into a sealed-off area which Dr Lambert announced as the sanctuary. The word held the promise of care and nurturing and protection but something told me that it would turn out to be a misnomer. We entered and he switched on the lights. I quickly adjusted my focus and took in the sight of several cages. Some had dogs: from large alsatians to small terriers. The cages were steel grids with enough space between the slats for excreta to pass through. There were, of course, no rugs or any type of comfort dear to the canine heart. They sat and stood and slept and lived in and on that steel grid. I stared at these cages, my heart beating fast, a vice-like spasm clutching at my chest.

No—I did not witness any frightful experiments and nor did I see any of the animals with tubes running through their bodies and sensors attached to their brains. Those sights are not open to the public. The scientists and research industry are careful to commit their sins behind an impenetrable iron curtain. But what I did see was a gathering of abject, miserable animals who had lost their freedom for ever and knew deep within their frightened little hearts that they were on death row and there was no running away.

I shall not debate the larger issues that surround

vivisection. The arguments are far too complex and there is no single approach that can be termed as the 'right' way to do biochemical testing. If there has been benefit to mankind through animal experimentation there has also been plenty of damage.

The tragedy of thalidomide comes to mind. A sleeping pill, thalidomide was released in the western market in the year 1957 after extensive tests on animals showed the drug to be completely safe for pregnant women. Apart from being a sleep-inducing drug it also relieves nausea and so, over a period of time, became a popular remedy for women who were suffering from morning sickness during their early months of pregnancy.

In fact, thalidomide was anything but safe. Within a matter of months the first generation of thalidomide babies began entering the world. The effects of the drug on the foetus had been horrifying. The babies were born with multiple birth defects: babies with stunted arms and legs, some with no limbs at all, some with deformed eyes and ears, other with ingrown genitals. Of course it was too late to reverse the situation, but the world was terrified and outraged and there was a sudden suspicion: what is the proof that drugs that are released after all the so-called elaborate testing on animals are, in fact, safe for human consumption? Animals differ radically in their fundamental physiological make up to man.' Therefore, it is very likely that drugs that have shown no side effects when tested on animals, may have drastically different results when used by humans.

The debate on vivisection continues with massive

propaganda being churned out from both the pro- and the anti- lobbies. But what I recall most vividly from my visit to the laboratory was a moment when I extended my hand to a haunted looking Basset hound. He recoiled, obviously having lost all trust in the human touch, and with his head thrust upwards let out a most dismal and painful howl. He continued howling for a long time. This set off a chain reaction with the other animals and soon all the cages were vibrating with the howls of those anguished little creatures. I rushed out of there and found myself in yet another laboratory with more animals in cages—this one had cats as well. The cats I saw were evidently newcomers who had not yet discovered the baseness of man. They arched their backs and rubbed themselves against the bars of the cages, purring loudly all the while. There were not many monkeys, they had almost all died, I was told dryly by Dr Lambert. One little monkey, however, sat in a cage similar to ones the dogs had. He sat there bent over like a little old man. One of his hands never stopped twitching and as I came nearer, he looked straight through me with unseeing eyes. His mind was elsewhere—or perhaps he no longer had one.

Next I saw cage after cage of rabbits. I knew what rabbits were used for. I had recently been shown classified pictures by a friend in the medical profession who was rebelling at what his job demanded of him. The pictures were graphic and tragic: of rabbits used for experimental purposes, their eyes burnt out by acid.

Further down the room, two small pregnant bitches eyed me distrustfully as I approached their cages. I wondered what tortures awaited them. I had heard of the wombs of pregnant bitches being sewn up to see how long they could remain in labour.

I walked out of that room with the sound of their howls ringing in my head and the vision of their misery and hopelessness swimming in my eyes.

Out in the corridor we ran into a junior assistant who was working late. He enquired as to the commotion. I accepted responsibility for throwing the animals into a state of agitation. He looked at me firmly and said, 'Well, Miss Rogers, you know these aren't your run-of-the-mill pooches and kitties. They are different and don't understand overtures of affection.'

I was shocked. Every animal that has been domesticated in any way knows what it means to be stroked. To be talked to. To be cuddled. Every warm-blooded animal has the ability to feel love, pain, depression and many of the same emotions that humans experience. They have the capacity for thought and memory as well and it scared the daylights out of me to imagine what this kind of incarceration would be doing to their psyches.

Can any of us just for one moment imagine what it might feel like to be confined in a cage about the size of a double bed, and of equal height, and to remain in it without ever coming out for the rest of one's life?

I was shortly dropped back to my lodgings. I went to bed but not to sleep. I lay thinking of those animals whose first heritage should be freedom, confined night and day to those gridwork prisons, with no hope of ever receiving release. To what low ebb has our moral sense deteriorated if our health has to be bought at the expense of torturing animals.

Perhaps some day, hopefully not too long in the future, the medical profession may be able to find alternatives to this ruthless practice. I quote from *Country Fair*, a periodical published in England, dated March 1970:

> **Replacement of Laboratory Animals**
>
> Techniques are now being developed to include the use of computers, tissue, cell and organ cultures and gas chromatography to find practical ways of replacing animals in many medical experiments, for more reliable and acceptable methods.
>
> So far scientists who have used such techniques claim them to be scientifically, economically and ethically superior to those involving animals, besides giving much speedier results.

Yes, it is true, now as we stand here in the 1990s much has changed. Yet we are still not out of the Dark Ages. Animals continue to be poached and hunted, slaughtered and skinned, and they continue to be used for medical research. What has changed is that more people in the world today are feeling accountable for the ruination of this planet and the decimation of hundreds of species of animals and birds. These are the brave souls who will eventually bring in a new dawn.

The day after my visit to the laboratory I had to leave America to return to England. I had a booking to read one of my poems at Westminster Abbey on the kind invitation of Mr Carpenter, the wife of the Archdeacon of the Abbey.

As I strode to the podium, poem in hand, I had the distinct feeling that I was going to faint. Thankfully I didn't, but as I read out the words the only vision in my eyes and the only sound in my mind was of the animals I had seen but a matter of hours ago in another country, far away.

We were now separated by a great distance and there was nothing I could do for them. They were voiceless pawns in the hands of the scientific world and mere mortals such as I could not change the course of their lives. Perhaps, they would go on to assist in the ongoing processes of science, to play their unfortunate roles in the discovery of new miracle drugs and miracle cures for humanity. But, let it be remembered, not a single one of us would ever recall the circumstances of their lives, the nature of their deaths or the value of their sacrifice. And that, to my mind, is a tragedy.

The cruelties heaped on animals in the countries of the East are more in the nature of man exploiting beast for basic survival reasons. The economic conditions of a vast majority of people in India necessitate a dependence on the beast. The *madari* depends upon monkeys for his daily sustenance. The *bhaluwala* depends on his bear to ensure that his two square meals come through. The snake charmer's income is generated by the dance of the cobra. The farmer in a backward area of the country, who has no access to modern means of agriculture, will rely on a sturdy pair of bullocks to till his fields without which his family might have to battle starvation.

To pass judgement in such circumstances would be unkind. This is a scenario where both man and beast are going through the hardships of life together. One would hope that the situation improves for all the creatures of God and until that happens all I can wish for is that the man who uses an animal as a means of income accords it the level of kindness and respect that it deserves.

Little things that run and quail,
And die in silence and despair!
Little things, that fight and fail,
And fall on sea, and earth, and air!
All trapped and frightened little things,
The mouse, the coney, hear our prayer!
As we forgive those done to us,
The lamb, the linnet and the hare
Forgive us all our trespasses,
Little creatures, everywhere!

James Stephens

Fifteen

HUSH, SAID THE DOCTOR

The trip to America and Canada in early 1970 was a success, though we did not raise nearly as much funds as had been predicted by Bo Goran and the lady medium. I visited several cities in the US—Washington, New York, Boston, Las Vegas, and was, as always, hosted by enthusiastic members of The Animals' Friend, who apparently considered it their holy duty to organize fund-raising events for me. After six months of dashing about from city to city and putting in maximum energy into the acquisition of cold cash for TAF, I found myself heading home again, with a tidy sum under my belt and hopes of expanding the shelter.

Coming home was always so heart-warming. Especially when I brought back small gifts for all and sundry. I never could afford anything fancy, but the boys (and their families) at the shelter were simple people for whom it seemed that the thought was always more valuable than the gift. A key chain with the Statue of Liberty painted on it, a bottle opener studded with fake rhinestones or even the shabbiest comb and hair brush set made of cheap plastic could send my troops into raptures.

During the time I had been away a certain amount of construction had already been completed at the shelter. We always went through these sporadic bursts of expansion and growth every time I went on a fund-raising mission. It was invariably assumed that the old lady would in all likelihood come back with pockets, if not laden, at least slightly bulging and that was justification enough for the shelter to feel wealthy and gung-ho! But the fact was that no matter how much I collected it was never enough. No amount of money was ever enough for us to achieve the kind of goals we had set for ourselves.

People would constantly corner me with allegations: Look at the stray dog problem! It's growing day by day. What are you doing about it?

The truth is, it is not easy to just waltz off one day, catch a stray dog and neuter or spay it in order to prevent it from furthering its tribe. No indeed, it is not easy. Every such act requires free-flowing funds. The spaying and neutering of dogs is a medical procedure and requires the fundamentals of surgical paraphernalia just as a human hysterectomy would require. The same people who screamed their lungs out about how ineffective animal organizations are, would scatter like nervous deer when requested to contribute toward the process.

*

What I came back to at the shelter was hardly a bed of roses. I was back in time for Christmas, an occasion we all looked forward to. I had taught the boys some Christmas carols and for several years we had been celebrating the silly season in each other's company. We would sing through the cold December night, repeating the same carols over and over again. And then with gay abandon we would gorge on a vegetarian feast specially prepared for the occasion.

Unfortunately, that Christmas was destined to be a dismal affair. We had an influenza epidemic at the shelter. Everyone was falling prey to the vicious disease. I spent my time ferrying people back and forth to hospitals, before finally succumbing myself.

Barely out of influenza, February saw us battling a hepatitis epidemic among the dogs. It was just wretched: practically all of our dogs died miserably and we could do so little about it.

In March again there was illness amongst the staff. I was ready to collapse. The last straw was when Vigin, not only my cook, but by now also my guide, philosopher and dear friend, was unable to leave her bed for ten days. I was truly very weary of the whole situation and, in fact, even remember thinking why I had chosen such a hard life for myself! I also had another personal problem at this time, which was intruding itself more and more forcibly on my peace of mind.

It was while I was in England in 1968 that I had met my old bank manager, recently retired, who felt it his duty to give me a kindly piece of advice.

'Miss Rogers,' he had said, 'don't you think it time you gave up this voluntary work of yours and turned to something more profitable?' My meagre income came through two basic sources: a pension and a social welfare stipend. Over the years, however, I had, like a wise old owl, been investing in shares as well, which had turned out to be reasonably profitable.

My inflamed response to his suggestion had been, 'Never!'

'But you know,' he complained weakly, 'you really can't go on selling out shares every time you need that little extra. At this rate you soon won't have any shares left.'

'Never mind,' I consoled him. 'Perhaps I'll die before the shares run out!'

'That's true enough,' said my banker friend, 'but you know very well that you can't really bank on dying!'

I had dismissed this conversation from my mind and had sold out shares several times since then but now it was becoming more and more apparent that I couldn't go on like this indefinitely.

What could I do to augment the income for both myself and the shelter? One thing was very clear, nothing could make me leave The Animals' Friend. What was it that Bo Goran had said?

'Are you doing any writing? You should write, you know. Perhaps a book on your experiences in India and about the animals you have cared for?'

He wasn't the first one to have said so. Over the years plenty of people had encouraged me to write and in fact, I had. Not that the writing had brought in much money. I had written mostly stories for children and some scattered articles for magazines on psychic phenomena. I just didn't have the time to write for myself. The kind of writing which is unhurried, where you have all the time in the world to delve back into the bottomless well of seven decades of experience and pull out all those delicious, hidden nuggets of life to share with a reader. How could one be so self-indulgent when the phone never stopped ringing, when the sheep never stopped bleating, when the dogs never stopped baying, when the monkeys never stopped chattering?

Besides, who was going to read the ramblings of Crystal Rogers, the so-called mad English memsahib extraordinaire? It would have to be an extremely patient reader with a deep affection for animals and one who could forego the glamour, action, and thrill of the current writing trends and accept my unpretentious life and old-fashioned prose.

Sometimes writing one single page for The Animals' Friend magazine could take up to an entire day because of the hundred and one interruptions. And if I did write, who was to know whether it would bring me even a penny? One had to be so brilliant and intellectual and original to get something published, and Heaven knows I was none of the above. Anyway, I told myself, nothing ventured, nothing won . . . So, I decided to take time off to marshal my thoughts and attempt to cobble together a string of recollections.

Sneaking off again was not as easy as I had imagined it might be. Dr Bhatia, who had been in charge of the shelter while I had been away on my last fund-raising trip, was determined not to be responsible again. He had encountered too many pitfalls and was not prepared to lose more sleep over the daily administration of the shelter. I could hardly blame him.

A slow resignation was seeping over me: if Dr Bhatia was unwilling to assume the responsibility of running the show I would have no option but to remain at the shelter—which meant no time, therefore no writing, therefore no book, therefore no (possible) extra money. It was saddening but not the end of the world: And then, quite out of the blue, John came along.

John Hardy was a friend of my friend Jill Buxton and a very talented and able young man, but at the moment his affairs were in a state of confusion and he was waiting for some of his future plans to materialize. Between the two of us we determined that the best course of action would be for John to bide his time at the shelter while his life sorted itself out.

So, John came and never had anyone fit in better. He threw himself into the work of the shelter with complete zeal and within a week's time knew the ins and outs of shelter life almost as well as Mowgli knew his jungle!

As luck would have it, the preceding months of sickness at the shelter and all the accompanying pressure had caused in me a condition that can best be described as extreme laryngitis. I must admit to feeling quite a wreck and looking it, too, I'm sure, but it was my voice that took the cake. Like a very under-the-weather crow I still managed to hop around coughing out instructions until the ENT specialist put an end to all of that by thrusting something that felt about the size of a machine gun down my throat. He proceeded to pronounce that I should rest my throat completely for the next few months!

How on earth could one achieve that in a shelter where one's time was spent perpetually bawling at animals or staff? For a few days I tried whispering. For some psychological reason everyone started whispering back and then I couldn't hear what they said. It was quite maddening!

The committee members insisted that I be packed off on a holiday. John volunteered to take care of everything at the shelter and I was quite confident that he would manage. I would get the desired rest for my larynx and, above all, here was the opportunity of a lifetime: free time to put feelings down on paper, to liberate locked-up memories, to get to know myself again after years of losing myself in others' lives and others' dreams. It was all too good to be true.

The end of April saw me on a bus making my way to Srinagar, Kashmir, accompanied by my dog Zena and Peter Firebrace, the teenage son of old friends of mine. I also carried with me a pretty new fountain pen that I had purchased especially for the purpose of giving life to long slumbering memories.

The first few weeks were spent with my dear old Kashmiri friend, Mr Mukoo, who laid out the red carpet for us. We spent some weeks on a houseboat soaking in the gentle warmth of the sun and the natural beauty of the surroundings. One morning, as I sat on the deck, reading a book, a dog of medium size, speckled brown and white, came and stood squarely opposite the houseboat on the bank of the lake. She appeared ownerless, a bit unkempt and undernourished. She stared at me. I gazed back at her. I could not take the staring contest any longer, so I strode toward her. She trembled, but stood her ground. Upon reaching her, I discovered one of her eyes was badly inflamed; it was half-shut and there was a rim of caked blood encircling it.

'Oh, you poor girl!' I exclaimed. The little creature got suddenly agitated by my voice. After a lot of coaxing and wheedling she allowed me to come closer and examine her eye. The wound was probably the result of a hard blow from a stone. I requested her to wait while I rushed in and got some teramycine ointment. She looked very bright and I was hopeful that she would understand the request.

She did. I cleaned her eye and applied the medicine and then handed her a biscuit which she chomped with great relish. She never moved while I doctored her eye, clearly realizing that I was only trying to help. 'Good girl!' I said, the dressing over. 'Come tomorrow and we will do it again.'

'Girl' evidently understood. The next morning she was there and rose to greet me with a wagging tail. From then on it became her daily routine: she was there everyday at the appointed hour to have herself treated and get her little biscuit treat. It gave me a great sense of satisfaction to watch her eye heal. The day came for me to leave the houseboat and amongst those on my farewell committee was 'Girl'. I patted her and she licked me as an expression of her gratitude. (As a post-script to this little incident I will add, dogs have a remarkable memory and tend to recall a good deed far more swiftly than humans. The following year I returned to Kashmir and found lodgings on the same houseboat which was anchored to the same spot. 'Girl' must have sniffed the news, for a day later she was there to meet me with a madly wagging tail.)

Next on the agenda was to move further up into the mountains to Pehalgam at which point Peter and I parted ways. I missed him dearly, he was wonderful company; but he had many places to see and much trekking to do and his youthful spirit needed to be set free, so off he went. This left me with plenty of time to actually begin the process of writing. I had settled comfortably into the farm belonging to a friend's friend and all that was left now was for inspiration to visit.

And soon enough, it did. The surroundings helped, no doubt—the crackling blue mountain sky, the gush of the cool morning breeze through tall pines and stout walnut trees. The peace of distant pastures dotted with grazing cattle and ponies. The music of rushing streams. The words came flowing out of me like endless ribbons. I could barely stop them; such was their volume and velocity. It didn't matter that perhaps no one would ever read them; it didn't matter that they might never find a slot. All that mattered to me was that I was reliving experiences and relishing them all over again. It was like having a projector inside that was spewing forth all those little moments that I was certain had faded with the passage of years. (Here, I must add a word about the book that I had put together. Some time later, I sent the manuscript to a friend of mine in England in the hope that he might find a publisher for it.

He mislaid it and for a couple of years there was no news as to its whereabouts. Well, I thought to myself, just my luck! However, one day I received a telegram from the same friend announcing: 'Your book found. Burglar came and ransacked house and found it.' To which I could send only one short reply: 'Please thank burglar!')

Shortly after the flood of words, something happened which threw me off course completely. I had heard through the local grapevine that the stray dogs of Srinagar and Pehalgam were being killed in the thousands by strychnine poisoning, a method already condemned by the Animal Welfare Board on account of its extreme cruelty.

Back in Delhi we had been receiving reports on all sides regarding the cruel killing of street dogs by municipal sweepers who went around throwing poison indiscriminately on the footpaths and near the rubbish bins. Often, along with the street dogs, pets belonging to good homes would also fall prey. After some time the four-footed nomads understood where the danger zones lay and instinctively avoided those areas. With a little urging and some cash incentives the sweepers were made to change their battle strategy: the new method was to roll the strychnine in a roti and entice the dog to eat it. This the trusting, hungry street dog would do only to die a few minutes later in dreadful agony. The ones who died quickly were the lucky ones, otherwise it could take hours before the poison brought release to a racked spasm-stricken little body. My frustration as an animal-lover was intensified when I realized that the poisoning was being done with the blessings and approval of the concerned authorities. Of course, being one of those who found it hard to accept the immutability of a situation, I decided that only by whipping up a public outcry could the problem come into focus and consequently get solved. There are many different ways to offer death and the authorities could surely find a more humane alternative—if they put their minds to it. In our monthly magazine of The Animals' Friend we began appealing to our readers to help us fight this cruelty in any way they could. By sending us full particulars of any case of poisoning they might have witnessed, by taking photographs, if possible, of the diabolical work in progress and by taking down the name of any sweeper actually seen administering the poison. The reader response was short-lived. The vile act continued despite the law against strychnine poisoning and dogs continued to die in agony day after day, month after month, year after year.

The disturbing news of the fatalities of street dogs in Kashmir once again forced me into action. Father James, a clergyman who was also a guest at the farm, joined me in a bit of detective work as I went around establishing the exact numbers of dogs that were meeting their end this way. Of course, exactitude is not something to be found that easily in India and we couldn't really get any hard figures.

One evening as I was out on a stroll, I heard a terrible, plaintive, anguished howl coming from further ahead. This was followed by complete silence. As I progressed on my way, I noticed a dark form writhing in the bushes that bordered the road. Parting the leaves I saw a strong-looking black bitch convulsing horribly. There was a trickle of fresh blood trickling down from her nose. Her eyeballs had turned all the way back, the limbs were thrashing about and her breathing was short and spasmodic. I watched her in horror, quite at a loss about what to do. As I gaped at her she took one final, furious turn, flipping onto the other side and with a last terror-stricken howl, she died. I could not be sure that she had been a victim of strychnine poisoning but the symptoms certainly seemed to indicate it.

Zena, who had accompanied me on this walk, stood at a distance, her small body shivering. She was too shocked to bark. She looked at me in complete bewilderment. I tried to console her on the loss of her fellow-being but the words got stuck in my throat as I tried to articulate for both of us the sight we had just witnessed.

I hurried back to the farm where I wrote a detailed letter to the municipal authorities, to bring to their notice that there were more humane ways to put an end to a dog's life.

My holiday was coming to an end and it was time to get back to the hustle and bustle of city life. The positive aspects of this trip were the recovery of my voice, the evolution of a supposed book and a letter requesting the discontinuation of strychnine poisoning in the municipalities of Srinagar and Pehalgam. Unfortunately, in all probability my letter was lying unopened amongst a horde of other similar envelopes in a blue, tattered old file on a wobbly old table in a musty old room in a crumbling old office . . .

Sixteen

GARAM MASALA!

I will now share a piece of information which I found quite amusing. My relatives and certain friends who lived in England and other countries in Europe were convinced that poor, dear Mishy must indeed be leading a miserable life, bleak and full of hardship. Quite the contrary. Hardship, yes. Bleak? Certainly not. My days were vastly more interesting than theirs! How could it not be? Imagine a household of thirty-odd humans and around seventy animals of assorted shapes, sizes and species. My life was as complicated and just as spicy as a rich blend of *garam masala*.

I will reconstruct the special flavour of that life by excerpting from the editorials I wrote for The Animals' Friend magazine. These span several years and do not necessarily appear in any chronological order:

> Since our last issue India and Pakistan have been at war and blackouts, food rationing and the various attendant evils which go with war, have been making us somewhat anxious for the future of our shelter . . . The recent cold spell, which despite all precautions, carried off two of our small puppies has been just as unkind to

us humans. Our electric supply chose just this time to fail us. With the heaters barely operating, the evenings have been unbelievably cold for all of us and my assorted cats who share my *razai* and have got considerably pampered by the warmth from the heaters! Recent arrivals to the shelter: three small puppies who were found crowded around the dead body of their mother. They are still in mourning. We shall organize their naming ceremony soon, which we hope will help to raise their spirits . . . We have had rather an influx of dry cows recently, put to pasture by their owners who find that they are quite useless to them if they can not deliver their prescribed ration of milk. Our fodder bills are assuming quite alarming proportions and any financial assistance would be much appreciated . . .

*

The monsoon is upon us once again and the low-lying areas of Delhi are flooded. The other night we were called on an emergency mission to rescue a terminally ill pony. We rushed out in our large ambulance and had to contend with pools of collected rain water, in places almost three feet deep. With a prayer on our lips we finally managed to reach the appointed address. Within minutes of our arrival the pony passed away. Perhaps, it was the best thing for him. However, we were soon to discover that our ambulance was also a patient. Its engine sounded very distressed and after it had barely covered a couple of kilometres it decided to get stuck in a large slick of wet mud. It took us hours to organize a police crane to pull the ailing ambulance out . . . On the happier side we now have two delightful babies in the shelter: Raju, a calf, and Dinny, Dinah the donkey's son—like all young donkeys he is a bewitching baby. Raju, who lost his mother in a road accident, sleeps on my bed, which is now getting considerably crowded. Three cats, Zena, my faithful dog, and now Raju the calf!

*

In recent weeks we have been called to the aid of two monkeys, both possible victims of animal experimentation. The first one came to us completely blind with empty eye sockets. A medical student in his third year who happened to be visiting us, saw signs of an operation having been performed on the monkey. The second monkey was picked up near one of Delhi's hospitals—a possible escapee. She is almost completely bald, with practically no teeth, which leads us to believe that she was being experimented upon for some form of vitamin deficiency.

*

One of the dullest animals we have ever had the opportunity of serving has been Alfred, the flying fox. Alfred winged his way into our shelter some nights ago and we discovered that he had an open wound on his left wing which needed to be treated. None of us has ever had any prior experience of ministering to a flying fox, so with a prayer we plunged into the act of medicating his wing. He was taken gingerly in a net to the medical room whereupon he was set free. He immediately found a comfortable perch on a curtain rod and flipped himself upside down. After that, it turned out that Alfred was actually very sensible. He remained motionless through the entire procedure. His wing improved day by day, but Alfred as a personality did little to endear himself to any one at the shelter. He remained hanging for the entire duration of his stay there, and then, one evening, realizing that he did not need us any more, he took wing. Quite an opportunist, that fellow!

*

The rains have been unforgiving this year. The shelter is practically flooded, which makes it a veritable paradise for Friday, our resident buffalo. The cats are most vocal in their objection to the water that surrounds them, and the humans are just as noisy. We have had a huge number of large animal cases coming to us, primarily road accidents: the roads, being wet and slippery, are not safe ground for animals crossing from side to side . . . the other day we got a call to attend to a young camel; she was covered from head to foot with warts the size of cauliflowers. Even her eyes were partially covered by these growths and the poor animal had a hard time moving around. Dr Bhatia, our veterinarian, is optimistic that she will recover soon. This suffering creature had plodded all the way from Rajasthan to Delhi, carrying hefty loads on her back. The fate that awaits her is the slaughterhouse, where camel meat is apparently a rarity. We have petitioned with the local authorities to save her life and others of her ilk. If things continue this way, between now, the late sixties, and the beginning of the next century, camels will be an endangered species in India. The generation of the new millennium will know of these remarkable beasts only through poetry, literature and documentaries. Yet another tragedy for this planet.

*

Talking about large animals, we got a telephone call some time ago. An agitated male voice enquired whether we could do something to help an elephant?

For a brief moment I wondered how on earth we were expected to transport an elephant, if it needed transportation. Camels were hard enough, but an elephant! We asked what help the elephant had already received and were told none. Many organizations had been approached but for one reason or another, all had managed to wriggle out of the responsibility.

The elephant was lying on the road near the Delhi-Punjab border. We reached there at about 10 p.m. the same night and found the elephant sprawled on the side of the road, near some half-built houses, in obvious misery. The elephant belonged to a small band of swamis. They told us that it had been involved in a road accident about a month ago. It was only five days ago that it had collapsed and was unable to rise again. The reason was not hard to find. Two of its legs were gangrenous and it was plain to see that it did not have long to live.

We asked the swamis to give us permission to put the

animal out of its misery. They would not hear of it, but added as an afterthought, 'If you would like to give us compensation of three thousand rupees, you can do as you like with it.'

We left, feeling quite frustrated and distressed, but returned the next morning to leave some water by the side of the elephant. It was time to find a magistrate who would give a court order which we could show to the swamis and finally ease the poor thing out of its misery. Being a weekend this was not easy, and it was not until a friend got in touch with the secretary to the Prime Minister that the wheels were set in motion.

It had come as a terrible shock to me that the swamis had not even bothered to administer the most basic kindness to the animal that had served them in the past. Not a drop of water! In five days! The words of Ella Wheeler Wilcox, the well-known humanitarian and animal lover, came back to me:

> So many faiths, so many creeds,
> So many paths that twist and wind,
> When just the art of being kind
> Is all this sad world really needs.

When the animal lay there cold and unmoving, I thought to myself: how the mighty fall. The elephant, such an awe-inspiring and majestic creature, but how vulnerable and pathetic when lying, slowly dying in the sun amidst the dust and flies on the Delhi-Punjab border.

Seventeen

TWENTY YEARS PACKED INTO A SUITCASE

The year that I will commence to describe is 1979. What one might call a watershed year for me. It was, with all due respects to Charles Dickens, definitely the worst of times.

I had been a resident of Delhi for twenty years, having first arrived in the capital in 1959. These two decades had exposed me to a panorama of fantastic experiences. Many had caused great heartache, but now as I look back, it is not with any sense of bitterness, but rather with the feeling that I learnt a lot more living through those moments of agony than I would have otherwise. However, one particular incident was to come as a crushing blow; even ten years later as I write this my eyes sting with hurt, my pulse races with outrage.

I was carrying a good feeling with me the previous few months: in 1978 I had been awarded the Robert Martin Award, the highest appreciation in the United Kingdom for work among animals. The award in itself was not something that made me dance with joy, but it did give The Animals' Friend greater credibility; it made us look more serious in the eyes of the world.

Friends were beginning to feel that I had essentially bitten off more than I could chew: running the shelter, editing and writing for both the junior and adult sections of the TAF magazine, driving the ambulance whenever the driver was ill or on leave and attending to the animals. It was, come to think of it, quite a handful for a seventy-three-year-old woman. It was suggested that I find a 'nice young man' to look after the donkey work and stand for presidentship of the committee.

I was not really keen to do this, but to have some of the responsibilities off my shoulders did present attractions. Without much difficulty the nice young man was found and I prepared to stand for elections for the post of president.

However, as word leaked out about me contesting for the post, a letter arrived signed K.K. Kumar, life member of The Animals' Friend. The letter pointed out that since Miss Crystal Rogers had come from the land of Asian hatred, namely South Africa, she was quite unfit to be president of this society. I recall laughing out aloud. It was preposterous! First of all, I did not hail from South Africa, I had been a mere visitor to the country in 1954. Secondly, what Mr K.K. Kumar obviously did not know was that I had worked with a most remarkable man by the name of Alan Paton, author of the thought-provoking book *Cry, The Beloved Country*, and had served as secretary of the Liberal Party, leading processions of 'non-whites'. I had written anti-apartheid poetry and composed revolutionary songs which incurred the displeasure of various officials and had also got arrested for my efforts.

He also was not aware of the fact that I had deep respect for Mahatma Gandhi and many of his ideals which certainly did not qualify me in any way as a 'white-supremacist'! Upon further investigation we discovered that there *was* no Mr K.K. Kumar, life member, and it was apparent that the letter was quite bogus.

However, this letter and its allegations fit in with the general mood at the shelter. For almost two years I had sensed a rift amongst the rank and file of The Animals' Friend. Small issues which could easily have been discussed and sorted out peaceably snowballed into major arguments. There were two clear factions that were evolving: one that was faithful to me personally and to my policies and the other that was against everything I stood for.

I had been attempting to shut my eyes to all this idiocy and had been quite successful in drowning myself in work, but it was clearly the attitude of an ostrich and even though (or perhaps because) I tried to ignore it, the situation merely worsened.

I was out on a fund-raising trip. Since around 1977 donations had reduced to a trickle. It took all my energy to drum up even the most meagre of donations. It surprised me that this was happening but there really was no time to sit back and analyse why people were suddenly turning even more miserly than before. I returned to the shelter to receive the shock of my life. Kundan, who had grown to be our veterinary assistant and 'compounder' (pharmacist) because of his utter devotion to the animals and his skill in managing their illnesses, had been sacked. This had been done by the committee without consulting me, without even waiting for me to come back! I was very angry and very upset. When I asked the reason for his dismissal I was told that he had been found drunk on his day of leave.

'Did he misbehave with anybody?'

'No,' came the reply.

'Did he cause any damage to either the shelter or to any of the animals?'

'No.'

'Is he a habitual drinker?' I knew for certain that Kundan never touched alcohol except on special festivals and because he was so unused to it, it went to his head rapidly.

'No,' came the expected reply again.

'Then what is the reason for his dismissal?' I shot back at the committee. 'Even a hardened criminal is given a chance to defend himself. Did Kundan get that chance?'

'No,' came the sheepish reply.

During the next few weeks plenty of malignant stories floated around about Kundan. I was also not spared by the same harsh tongues and deranged minds, and plenty of hostile propaganda was churned out against me. I was very sorry for Kundan, for his abject humiliation, for the fact that the animals missed his magic touch. He had always been a model worker and I really couldn't believe that this was happening to him. When I tried to reinstate him, the committee informed me that he had committed a misdemeanour and bringing him back would have a negative effect upon discipline in the shelter.

This was the first open confrontation between me and the 'enemies' who had mushroomed around me.

I stood for election. The aftermath of the Kundan incident still hung heavy in the air. I was steeling myself for the very worst. I could sense the tension and knew that it would be a tough contest.

The counting done, the results were announced. Mrs Wahi, our President for the past seventeen years, was elected over me, and continued in her role. My defeat was complete.

Obviously, over the years the numbers had swelled against me. I was accused of being an inept administrator. I was accused of not allowing The Animals' Friend to develop into a profit-making organization. (That was against the very principal of charity work. A proposal had come from other committee members to reduce the number of sick and injured animals and instead build kennels for housing the pedigreed pets of the rich and famous while the owners went out on jaunts. And, quite naturally, to charge them the earth and moon for that service.)

It is odd, very, very odd, I thought to myself one night as I lay in bed, after this whole election fiasco. Here I am, the old coot who has slaved over this mission. I had spent the past twenty years of my life constructing an organization which I wanted to be run democratically. Creating a home for animals who had no homes, giving hope to people who had run out of hope, digging into my own funds to see that the animals were fed, the staff was paid, the shelter remained up and running . . . Yet, here I am today all alone, friendless so to speak, at the ripe old age of seventy-three, at the kind of cross-road that I would never have imagined I would arrive at.

My sense of having been betrayed was severe. I tried to gloss over the deep hurt this had caused me, but those close to me knew how much it had pained me.

As a peace-offering I was tempted with the position of the patron of the society. This was purely a titular role where I would have no power to act or to control in any way the organization I had so lovingly created. I must confess this sop outraged me terribly. It was the last straw as far as I was concerned.

After debating some more over the entire situation I decided to call it a day. To hand over the reigns to the committee as gracefully as I could …

It had never occurred to me to entwine the name of the organization with my own name. It had never occurred to me to promote myself in any way. It had always been more important to ensure that the organization have a life of its own, independent of me, that the work continue regardless of who manned the show. So, falling upon all my old policies and wisdom, putting aside the hurt for the moment, I announced to the committee that I would like to resign from all responsibilities associated with The Animals' Friend.

Those were some of the hardest words I have ever had to utter. My lips were dry while my eyes were moist. My insides churned. My head throbbed violently. My emotions had a life of their own, while outwardly I tried my best to remain calm for the minute or so that it took for me to state my case and for the roomful of members to murmur their acknowledgement.

And so that was it. Quite simple, really. Two decades of labour and love and duty and conscience. Twenty years of my life. And before I knew it, I was wrapping all of it into a suitcase with old cardigans, skirts, towels and shoes.

I had little notion of where to go or what to do. My only wealth in the world was the goodwill I had developed over the years with people I knew to be genuine friends. I prayed hard that someone would come up with a solution for me. Funnily enough, even though I had friends whom I could stay with, I was essentially homeless. As homeless as the sad-eyed, hollow-spirited people I had picked up off the streets and gutters and bus-stations of Delhi. At least they had had The Animals' Friend to go to. I didn't even have that option.

Besides, I was sure from now on the doors of the shelter would not be opened for humanity. I hoped they would at least remain open for animals.

And so, it was time to shake the dust of The Animals' Friend off my feet. Time to move on to better days. Or perhaps worse. Who could say?

Eighteen

WAITING FOR SUNSET

Actually, it did not take me too long to resurface. Living with animals one understands the whole meaning of licking one's wounds. You lick for as long as it takes a scab to form and then you are on the prowl again.

One of the first things I did was to re-employ Kundan. I could not imagine starting off any new venture without him for two primary reasons: the first, his genuine competence with animals and the second, I truly believed an injustice had been done which had to be righted.

I was quite sure that it was now time to leave Delhi. To approach anyone in Delhi with a proposal to launch a new shelter with me would have been awkward for all concerned.

I made contact with some friends in Jaipur, who were senior officials in the University of Rajasthan. It was decided between us that an animal shelter was badly needed in Jaipur and that I should be the one to set it up. So, bidding my farewell to Delhi, I set off for the dust and heat of the Pink City.

I pushed all the bitterness of the past weeks into a dark corner of my mind and focused on the prospects of a new life, a new city and new relationships.

The early days were not too bad, thanks largely to the fact that we were almost immediately donated land by the Rajmata, the former Maharani of Jaipur. An old dairy farm with its surrounding land became the base of operations. We established a charitable trust and I christened it Help In Suffering, for short, HIS. Once again, our mission was the same—to help suffering animals. Except this time I made sure we added the words: 'and all living things'.

So with my core staff, a small committee, a few injured and sick animals and a couple of human derelicts, Help In Suffering sprang to life.

Apart from some tumble-down stables, the dairy farm offered no amenities for housing either animals or human beings. My appeals to the Rajmata to be allowed to build on the land met with firm refusals. All I was given permission to do was to resurrect some old, fallen structures that were scattered on the premises.

Once again I went through the entire chaotic cycle of having cats in the bedroom, dogs in the dining room, hens in the kitchen, old ladies in the larder and so forth! As much as we tried to adjust, it was clear that we needed to expand and that this situation could not continue; however, for her own reasons, the Maharani remained adamant about not allowing any kind of construction on the property. So we were off on a land-hunt again.

Jaipur, being an arid area, did not provide many verdant landscapes and water was a perpetual problem. We were shown plenty of sights but none appealed to us. Meanwhile, the Rajmata was getting restive and wanted her land back. The committee of HIS was also showing signs of displeasure at my refusal of every site. I accept as one of my failings my stubbornness: it has been known to offend and annoy people. On the other hand the same mulish behaviour has been appreciated when someone has needed me to get something done. However, with my Delhi experience fresh in my memory, I did not want to upset the committee any more than I already had and resolved to move to the very next site that came up.

As it turned out, the land which eventually came our way was a kilometre away from the dairy farm and was perhaps the greenest, prettiest area in all of Jaipur! Naboth's Vineyard was its name and I thought it was completely delightful. I started planting flowering creepers on all the boundary walls, trees were plentiful and shady, and the animals took to it like it had been their home for ever.

We started building as soon as we got our building permit. We had barely started with a shack for a dispensary when I discovered that our funds were running out. I was so used to this type of emergency by now that the panic and fear that used to grab me in the early days had vanished almost completely; it was only a question of approaching the right sources and then hoping that help would come in time.

It did. Mrs Jeannine Vogler, a Swiss animal lover, heard about us and the difficulties we were facing and flew over to help. Finally, with the infusion of the funds that she was able to arrange, the shelter became habitable. Jeannine even sent over a Swiss couple to help with the work.

The couple had been informed that as long as they remained, they were to be completely in charge of the running of the shelter. I don't know how this came about, but at that stage I did not want to trample on anyone's toes, nor did I want to show a lack of gratitude for the donations that were coming in, so I decided to keep out of the picture as best as I could!

However, one very hot night, I went down to the kennels and found the dogs panting and looking miserable. The temperature must have been around 45 degrees. The earth was burning, I could feel it through the soles of my shoes. The Swiss 'manager' walked in on me. I pointed out to him that Jeannine and I had decided that during the summer the kennels must be equipped with fans—one at each end of the kennel.

He grunted. Shouting out for Kundan, he ordered him to get fans the next morning. Unfortunately, by the next day it was already too late. Two of the bitches had died of heat during the course of the night. I kept quiet and did not even so much as discuss the issue with the two managers.

This episode, however, was enough to make them write to Jeannine, saying that Miss Rogers was interfering with the work and making things very difficult for them. Sometimes, try as one might, one can't do anything right! To prevent any further bad blood I left the new shelter for as long as it took them to decide that they had had enough, and returned only when they were well and truly gone.

*

The years rolled by. It was business as usual. All the various ups and downs of life, similar to the shelters of The Animals' Friend. The same animal problems to deal with, the same human hostilities to contend with, the same joys and similar frustrations. There was one substantial difference, though. Every year was ensuring that I was inevitably older and feebler by twelve months. My vision was failing. My memory was not what it used to be. A lot of my earlier stamina had gushed out of me like air out of a balloon. I was turning into the old race horse now due for retirement. And I realized that many of the people around me—certain members of the staff, some of the members of the committee, a few people who claimed to be 'friends'—were beginning to recognize the symptoms of old age. And as such, I'm sorry to say, were taking advantage of me.

There were numerous occasions when I would ask for something to be done and it would not get done. When I would enquire about it, I would be told: 'But you never mentioned that.' In my presence it would be discussed, with accompanied sniggers of, 'Miss Rogers just can't remember anything right these days!'

Of course it was never quite that bad: my memory was diminishing but it had not quite been obliterated, as every one else made it out to be! However, I thought it best to play along and not take a confrontational stand on every issue. The months passed, the seasons changed and I grew older. And older. I tried to focus only on my work with the animals, distancing myself as much as I could from the usual complications of administration and committee functions. But destiny is a strange thing—much as one might like to keep trouble at bay, when it is time, trouble will seek you out even if you have hidden yourself inside a cave in an unexplored area of the Arctic Circle.

We were in need of money again. A son of dear friends of mine was with me at the time, helping out at the shelter. I liked the young man and trusted him as if he were my own. One morning, unable to go to the bank myself, I requested him to credit a cheque for me. It was a very large sum of money required for the shelter. I wrote out the amount and signed the cheque. He took it from me and in the most natural voice assured me he would be back in no time with the money. He stepped out and I returned to writing a letter to a friend in which I mentioned, 'Have just sent for more funds for the shelter. Thankfully S—— is here and have sent him to the bank. He has shaped up well. A dependable and likeable boy and so committed to animals. I wish more children around the world were like him.'

The morning stretched into afternoon. I ate a distracted lunch. The young man, S——, had not returned with the money. Afternoon dissolved into evening. I made a private call to a friend and shared my fears with her. She reassured me that there would be a good reason for his delay. I was not so convinced about that any longer. The night passed and S—— did not come.

He was not to be found. The money was not from my personal account—it was money that belonged to the shelter, to the committee, and the members had every right to know where it had gone. I undertook another few days of investigation and inquired about the boy's whereabouts before announcing to the committee that the money had been stolen.

The reaction was predictable. I was ready for it. There was pandemonium. The shelter funds had been stolen! By the boy that Miss Rogers trusted!

Strident voices rose, 'We knew this day would come.'

'She is becoming incompetent!'

'Her judgement is failing!'

'She will do more harm than good for the shelter . . . no, no. This can not carry on!'

Off with her head! Off with her head! The words danced in my mind. For everyone else it was an opportunity to unify and direct their collective rage against me. I understood the rage and even though it was humiliating for me, I sat through all of it with my eyes downcast. Their anger would subside with time but I was the one who was truly cursed. I would have to carry an unenviable burden of guilt for the rest of my life—yes, I had made a mistake. I had made the mistake, yet again, of trusting my fellow man.

This incident, in a way, truly broke my spine. Perhaps they were right after all, perhaps I was too old to be running shelters and making decisions and running around attending to animals. Maybe it was, after all, time for me to pick up knitting needles and find a nice old rocking chair for myself. But it was hard to convince myself that my old engine had run out of steam, especially since I still felt that, no matter what, my dedication to the cause was much stronger than that of most others around me.

However, my health was playing terrible tricks: over the years numerous broken bones, repeated attacks of pneumonia, and most recently a strangulated hernia caused me to seriously consider retirement. The knocks had been plentiful in life: how much battering can one's spirit withstand? Coupled with that was the doctor's recommendation: 'You cannot take these extremes of climate. The heat of the northern plains will take your life.'

I handed over the running of HIS to Christine Townend, an Australian girl who had assisted me in managing the shelter. Christine was a very competent and dedicated person and I was confident that she would run a tight ship.

It was time for me to put Jaipur and Help In Suffering, and the time that I had devoted to setting up the shelter, behind me and to move on. As I have mentioned before, there are many unhappy memories that I consciously wish to block out, and I shall not dwell more in these writings on those eleven long years in Jaipur.

Like the eternal wanderer I packed up my belongings and planned a final shift to Bangalore. For some of us change, displacement and uprooting becomes a way of life. On its brighter side this kind of lifestyle offered me the opportunity to walk away from unhappy memories, not that those memories ever really detached themselves completely, just that they didn't hound me all of my waking hours. On the dark side, I lost friends: people floated in and floated out of my life like wraiths. Even those I wanted to hold on to forever slipped through my fingers like spring water. All part of the game, I have told myself philosophically a hundred times.

Bangalore, I knew, would be my final move and there was a certain comfort in that realization. I was eighty-four at the time I arrived in this city. How many more years had the good Lord in mind for me? After all, I had swallowed much in the course of my life but had so far not had the honour of downing the nectar of eternal life! Mortals come and mortals must go . . .

The climate of Bangalore was pleasant and I found the people friendly. But the plight of animals remained uniformly pathetic across the length and breadth of India. I was taken in by Suparna Ganguly and her husband. Suparna, who I had known as a college girl in Delhi, was now settled in Bangalore with two young daughters. She and her husband, Kalyan, offered to help establish me on my doddering feet. Without them I should have found it very difficult indeed to settle into a completely new part of India.

Some friends had advised that I return to England. I couldn't even think of it. India is a strange phenomenon. It settles over one as soundlessly as mountain mists, as invisible as evening dust. The joy of India is in a bite of a firm, green guava sprinkled with rock salt. The unintelligible cries of the *sabziwala* as he trundles his cart over potholed streets. The chaos of babblers and sparrows as they wake up entire colonies on sweaty mornings. The sight of shiny, brown-skinned boys jumping into a canal for an afternoon swim. The pungent smell of fires burning on street corners on cold evenings, where men gather round and gossip about the day's happenings.

I could not and would not leave India. I have loved it with all my heart. I have been welcomed by the country and its people without hesitation. Yes, of course, there is much to criticize and condemn, but there is much more to appreciate and treasure.

I was biding my time as Suparna and others hunted for a house for me. On a particular visit to Sai Baba's Ashram in Whitefield, I noticed, as we drove out of the Ashram precincts, two men on a motorcycle whizzing past us. The one at the back was carrying a bundle of fowls, hanging upside down with their legs tied tightly together. As the cyclist swerved the birds' heads brushed the ground.

I was filled with the uncharitable desire to tie the two men's ankles together and give them the same treatment, hanging them upside down from an aeroplane.

The mercury was rising. The old flame flickered anew. Gentle visions of knitting needles and rocking chairs and an old biddy passing the evening of her life waiting for the Grim Reaper to escort her to the Golden Stair were dismissed. It was time again for action!

Suparna and I drew up plans for yet another animal welfare organization—to be named Compassion Unlimited Plus Action (CUPA). Suparna had many friends in Bangalore and soon we had a committee going with herself as honorary secretary and myself as president. My past experiences in Delhi and Jaipur had subdued me, no doubt, but not to the extent that I should never raise my head again. Twice bitten but only a little bit shy sounds more appropriate in my case …

We got going, with CUPA functioning out of my

house, though I had to move thrice before finally moving out of Bangalore to the suburb of Whitefield. Each move was motivated by complaints from landlords or neighbours wailing about the kittens and puppies, owls and crows that they had to share their premises with!

Whitefield is about twenty kilometres from Bangalore City station. It is almost pastoral, the population is low and the air is still clean. The house had a small compound but it was large enough to erect kennels and a wire shed for the puppies. The cats were kept inside the house in the main bedroom.

I was soon to discover that there was something fairly sinister about the house. The first incident occurred when I was sitting quietly, reading in the living room. A vessel containing water flashed past me, spilling water over me as it passed! It fell to the floor with a tinny sound. When I got up to recover it I found nothing there. Simply nothing! I could say that I had just dreamt it up, except for the fact that my clothes were dripping wet.

After this the cat room became the centre of strange manifestations. The kittens began to vanish. The doors always remained shut, the windows were bolted—but the kittens nonetheless managed to walk through the walls.

One of the most heartbreaking and inexplicable events was when two of my most favourite kittens lost their lives in very mysterious circumstances. I had some urgent typing to do which I couldn't possibly have achieved while juggling those two little tornadoes in both hands. I went into the cat room and put them down, shutting the door behind me. Five minutes later I looked in on them again.

They were fast asleep. My typing done, about an hour later, I returned to bring them out. I opened the door and found to my complete horror that one of them was lying dead in the middle of the room with its head bent right back as though someone had wrenched it fiercely; the other one was missing and was never found again.

It was more than I could take. I was prepared to personally confront a murderous poltergeist, if there be such a thing, but there was no way that I would allow the animals to continue to live in an environment which was, for whatever reasons, dangerous for them.

I moved a further three kilometres out of Whitefield to a village called Kadugodi.

Here I now live in a dream come true. I am in a cottage surrounded by swaying palms and green grass. I have with me four dogs and two cats. I share the premises with a training school where rural women learn weaving, sewing and even typing.

I have finally relaxed my pace. The flesh is indeed growing weak, as are the eyes. The committee at CUPA is a strong one comprising dedicated people who are today more capable than I am of ensuring that the battle for the weak, voiceless, four-footed denizens of this planet be fought and they be granted justice.

After establishing three separate organizations in three different cities of India, after labouring long days and nights, after writing down my feelings under a battery of assumed names from 'Karuna' to 'Billor', I realize again that it is not important whether I am remembered as Crystal Rogers alias Mishy, a champion of animal rights, or whether I am remembered as the *pagal* memsahib, or whether I am remembered at all. The only thing of importance is that the work should continue. I am thankful for people like Suparna Ganguly and the others at CUPA, Maneka Gandhi, CUPA's chief patron, and a host of individuals and organizations around the country who will take this crusade into the next century.

I will not be around to see the changes, but I am sure that the new century will bring with it a new understanding. New attitudes and a new generation of more humane humans . . .

*

I shut my eyes and see the little girl I knew myself to be, eighty years ago, a little girl running through a meadow, wearing a pink pinafore, racing barefoot down a slope after a bunch of colourful, fluttering butterflies. Not to catch them. Not to break their wings. Not to preserve them in a jar to show off to friends.

I know now that the girl was chasing after them to share their freedom. To celebrate with them in the divine, cosmic beauty of nature. To be a part of the exquisite balance manifest on this planet, where every living creature is granted its right to live and breathe and walk in dignity.

My prayer, now as I live out the evening of my life, is for all of humanity to share the same goal: to guarantee peace on earth. Harmony between man and man. And man and beast.

One of these days soon I will reunite with my dear friend Jim, with my family, my dogs, my cats, my goats and many other friends. They will ask me how I lived my life on earth. I will answer: I woke up with the song of the birds. I cared for creatures that could not speak. I fought with those who did not defend the rights of the defenceless. I tried to do the best I could.

I believe that the spirit lives on. I will die but my dreams will continue. They will find their voice through other people. They will find fulfilment through action.

Now I wait here, at my desk. I am waiting for the sun to set. From this window it is especially beautiful. The dogs are outside, growling at shadows. The cats are content and asleep near me. I remember a poem I wrote a long time ago, in the 1950s:

You say you love this warm kind world,
This human life you find so sweet—
'Tis not so hard to love the path
With flowers growing round your feet.
I do not ask you to forgo
One happy hour of one glad day,
The world and life and youth are yours
Enjoy these treasures while you may.
One day when you and I are old
And all life's joys are on the wane
The things you look upon as loss
To me will but appear as gain.
So clasp earth's treasures to your breast
And drink the cup of life as wine,
Yours is the present and the past
But all eternity I claim as mine.